KT-221-341

Analysis of Groundwater Flow

A. J. Raudkivi Ph.D., Dipl. Ing.(Hons), F.I.C.E., F.N.Z.I.E.
Professor of Civil Engineering, University of Auckland

R. A. Callander Ph.D., B.E., F.N.Z.I.E.
Associate Professor of Civil Engineering, University of Auckland

First published in 1976 by
Edward Arnold (Publishers) Ltd
25 Hill Street, London W1X 8LL

Boards edition: ISBN 0 7131 3359 7
Paper edition: ISBN 0 7131 3364 3

Set by Preface Ltd, Salisbury, Wilts.
and printed in Great Britain by Unwin Brothers Ltd,
The Gresham Press, Old Woking, Surrey.

Preface

The aim of this monograph is to introduce the student to the analysis of groundwater flow. The broad field of the physics of flow through porous media is surveyed briefly to give a perspective view of the problem, but the bulk of the book is devoted to simple analytical techniques.

The emphasis is on the potential flow analogy, which is based on simple concepts, and the use of more advanced mathematical techniques is avoided. By keeping the mathematics as simple as possible we hope to convey the logic of the approach to readers who are not too confident in mathematics. Thus, although complex variable methods are widely used in practical applications we have not included them. Neither have we dealt with the application of computers to solving groundwater flows. It is hoped that, by studying the logic and the advanced techniques separately, the reader will more readily come to grips with both. Our objective is to help the reader to understand the flow of groundwater and to apply this understanding to real flows. We have tried to steer a course between a text on higher mathematics on the one hand and a handbook of formulae and methods on the other.

It follows that the scope of our coverage must inevitably be limited. We will not embark on application of the logic to problems which cannot be solved without the more advanced techniques, nor do we attempt to cover all aspects of flow through porous media. There is an extensive body of literature on the subject to which we will refer and we hope that this book will serve as a stepping stone for engineering and other students who wish to read these advanced texts. Familiarity with basic fluid mechanics concepts and equations is assumed; however some readers may find our *Advanced Fluid Mechanics – an Introduction*, Arnold 1975, helpful.

A.J.R.
R.A.C.
University of Auckland, New Zealand 1975

Acknowledgements

The publisher's thanks are due to the following for permission to reproduce copyright material:

Council of the Institution of Mechanical Engineers for a figure from an article by H. E. Rose in *Proceedings Paper* 1945; United States Department of the Interior and the Bureau of Reclamation for a Table from Engineering Monograph No. 31; Pergamon Press Ltd., Oxford for an extract and a figure from A. J. Raudkivi's *Loose Boundary Hydraulics*, 2/e, 1975; Centrale Organisatie TNO for a Table from TNO, *Prac. and Inform*, No. 10, The Hague, 1964; Society of Mining Engineers of AIME for a figure from M. C. Leverett's 'Capillary Behaviour in Porous Solids', *Transactions*, Vol. 142, AIME, N.Y., 1941; Colorado State University for two figures from R. H. Brooks and A. T. Corey's 'Hydraulic properties of porous media', *Hydrology Papers*, Paper 3, 1964; American Elsevier Publishing Company Inc. for a figure from J. Bear's *Dynamics of Fluids in Porous Media*; Purdue University, Water Resources Research Center, Layfayette, Indiana for two figures from R. W. Skaggs, *et al.* 'An approximate method for determining the hydraulic conductivity function of an unsaturated soil'; Journal of Meteorology for a figure from J. R. Philip's article in *J. Meteorology*, 14, 1957; American Society of Civil Engineers for a figure from R. H. Brooks and A. T. Corey's article in ASCE *Proc.* (*J. Irrigation and Drainage Div.*), Vol. 92, IR2, 1966.

Units

Almost all formulae in this book are dimensionally homogeneous. As such, they are valid for any coherent system of units and provided the data used to solve a problem are converted to coherent units before substitution, these formulae will give results in the appropriate units of the same coherent system.

In numerical work, SI units have been used. These units comprise the metric system which is in use or is being introduced in many countries (including many countries in the British Commonwealth) and is likely to become the system of measurement throughout the world. In this system, the fundamental units of mass, length and time are the kilogram (kg), metre (m) and second (s). Corresponding derived units that appear frequently in this text are the newton (N) for force and the pascal (Pa) for pressure ($1 \text{ Pa} = 1 \text{ Nm}^{-2}$). Abbreviations signifying powers of ten are used as recommended by the International Organization for Standardization (ISO). Thus, 1 MPa means 1 megapascal or 10^6 Pa and 1 mm means 1 millimetre or 10^{-3} m. The following table shows a selection of prefixes likely to be encountered in this field:

giga	mega	kilo	centi	milli	micro
G	M	k	c	m	μ
10^9	10^6	10^3	10^{-2}	10^{-3}	10^{-6}

Conversion of units may be required before applying a formula to a problem in practice and we recommend the Stroud convention, (Walshaw (1964)). The permeability K is a variable which is particularly likely to call for such manipulation, since it is often measured in units which belong to no coherent set. For example, if K is given as 1200 gallons per day per square foot, conversion to the SI units ms^{-1} would be required before use in a homogeneous formula and would be achieved as follows:

$$1 \text{ gallon} = 4.54 \times 10^{-3} \text{ m}^3 \rightarrow \frac{4.54 \times 10^{-3} \text{ m}^3}{\text{gal}} = 1$$

$$1 \text{ day} = 8.64 \times 10^4 \text{ s} \rightarrow \frac{\text{day}}{8.64 \times 10^4 \text{ s}} = 1$$

$$1 \text{ ft} = 0.305 \text{ m} \rightarrow \frac{\text{ft}}{0.305 \text{ m}} = 1$$

Hence

$$1200\,\frac{\text{gal}}{\text{day ft}^2} = 1200\,\frac{\text{gal}}{\text{day ft}^2} \times \frac{4.54 \times 10^{-3}\ \text{m}^3}{\text{gal}} \times \frac{\text{day}}{8.64 \times 10^4\ \text{s}} \times \frac{\text{ft}^2}{0.305^2\ \text{m}^2}$$

$$= \frac{1200 \times 4.54 \times 10^{-3}}{8.64 \times 10^4 \times 0.305^2}\ \text{ms}^{-1}$$

$$= 6.78 \times 10^{-4}\ \text{ms}^{-1}$$

Difficulties with units can be avoided by strict adherence to the rule that coherent units and no others be used in dimensionally homogeneous formulae. In non-homogeneous formulae units must be used as specified.

List of symbols

A	area, constant
a	constant, rate of rainfall, specific internal area (a_s) of voids
B	constant
b	thickness of aquifer, constant
C	constant
c	concentration, constant
D	diffusivity, with subscripts for molecular, liquid, vapour, longitudinal and transverse diffusivity
d	particle diameter, depth
E	modulus of elasticity
e	partial vapour pressure, eccentricity
f	function, friction factor, rate of infiltration, Δf_s – free energy of soil moisture
G	dimensionless function
g	gravitational acceleration
H	potential (level) difference, electric field, H_r – relative humidity
h	piezometric head, with subscripts e for extremity, r at radius r, o of undisturbed level, w at well or drain.
I	integral, I_0 modified Bessel function of order zero
i	unit vector
J	Leverett function, Jacobian, J_0 Bessel function of the first kind of zero order
j	unit vector
K	coefficient of permeability [L/T] or hydraulic conductivity, K_0 modified Bessel function of the second kind and zero order
k	permeability [L^2], unit vector
L	length, L_e latent heat of evaporation
l	length
m	packing factor, strength of sink or source $= q/2\pi$, a parameter $= r^2S/4Tt$, exponent
N	constant
n	porosity, a variable, constant
P	scalar overburden pressure, chemical potential, rate of rainfall
p	pressure, with subscripts for capillary, bubbling, gas, liquid pressure

Q	flow rate, flow rate per unit width in two dimensional flow in vertical planes
q	flow rate per unit thickness in two-dimensional flow or radial flow, velocity vector
R	radius of sphere, gas constant
r	radius, with subscript e at extremity, w at well face
S	slope, storage coefficient
s	degree of saturation equal to the ratio of volume of water to volume of voids, drawdown, direction
T	temperature, transmissivity of aquifer
t	time
U	mean velocity in x-direction
u	local mean velocity component in x-direction, a dummy variable
V	mean velocity
$\forall$	volume
v	local velocity component in y-direction
w	local velocity component in z-direction
x, y, z	coordinates
z	valency, depth of flow
α	angle, shape factor, function, reciprocal of modulus of elasticity of solids
β	function, angle, reciprocal of modulus of elasticity of water
γ	specific weight
δ	distance of outer Helmholtz plane from crystal surface
ϵ	dielectric constant
ζ	zeta potential
θ	volumetric moisture content equal to the ratio of volume of liquid to bulk volume, angle
κ	the reciprocal of the thickness of the double layer
λ	parameter
μ	viscosity
ν	kinematic viscosity
ρ	density
σ	surface tension, normal stress, standard deviation (σ^2 – variance)
τ	period, parameter
ϕ	potential, potential function, function
Ψ	capillary potential or matrix suction head
ψ	stream function
∂	partial derivative

Contents

1
Introduction

The general topic of flow through porous media covers a wide field ranging from industrial processes in factories to the movement of oil or gas in an oil-field. In this book we are concerned with one aspect only, the flow of water into and through porous materials of the earth's crust. We begin with a general survey of the problems and the physical processes involved

1.1 Groundwater

Groundwater is all interstitial water below the water table, however deep down it may occur. The *water table* is the upper surface of the completely saturated ground, above which is the *zone of aeration* or the zone of suspended water which extends to the ground surface. This latter zone contains water stored as attached films on the surfaces of the particles, as well as water moving down by gravity (seepage or gravity water). The capacity of the soil to hold suspended capillary water and water in the attached films is called field capacity and it is the excess over the field capacity which is free to travel downwards as gravity or seepage water. The top layer of the zone of aeration contains a belt of soil (about 3 m thick) where the most active root development occurs. The zone of aeration is also of great interest to the geologist because it is there that destructive chemical action and weathering of rocks occur.

There is also a capillary fringe, which extends above the water table into the zone of aeration and where the water is held above the water table by capillary forces. The height of the fringe depends on the pore size and can vary from less than 1 cm to several metres. The maximum rise in soils is seldom more than 3 m. Figure 1.1 illustrates the meaning of some of this terminology and shows how groundwater can be classified according to occurrence.

The motion of groundwater can be subdivided as follows: seepage; capillary rise and capillary flow; percolation, which occurs only in the saturated zone of the porous medium, under the action of a hydraulic gradient; turbulent groundwater flow which may occur in large openings, for example, underground rivers. To these should be added movement of water vapour which fills the voids in the soil and moves along the pressure gradient as well as temperature gradient. Condensation of this water vapour can lead to formation of groundwater.

The influent water from the ground surface at first diffuses slowly downwards,

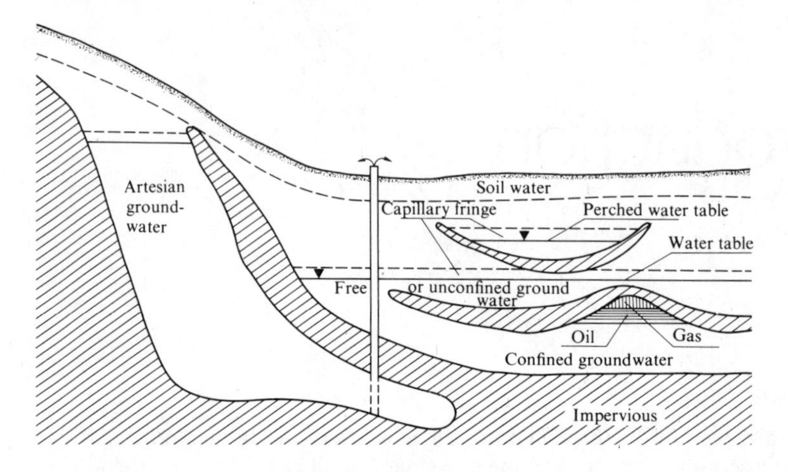

Fig. 1.1. Illustration of groundwater terminology.

wetting all the surfaces and replenishing the films on the particles. When all surfaces are wet, the movement of water is a seepage or gravity flow. This movement is complicated by the presence of ground air, most of which is expelled from the ground or dissolved in the seepage flow. Some, however, remains trapped in the soil as bubbles which reduce its effective porosity.

The influent seepage velocity consists, to a first approximation, of two terms: movement down the moisture gradient (a diffusion term) and seepage through a porous medium of varying permeability under a pressure gradient. Diffusion of the influent water is controlled by the moisture gradient and a diffusion coefficient. The first of these varies in the vertical direction and depends on the pressure, the suction or pore pressure deficiency of an unsaturated soil. The diffusion coefficient is a function of the moisture content and has to be determined experimentally. Percolation (in this case the downward gravity flow) depends on the pressure gradient and the permeability of the soil. The permeability, which will be discussed more fully later, is a measure of the ability of a soil to conduct water under the action of a pressure gradient. Like the diffusion coefficient, it depends on the moisture content and must be determined by experiment. When the gravity water has drained away, each grain retains on its surface a film of water which is in dynamic equilibrium.

Like any other fluid motion, flow through a porous medium has to satisfy the continuity requirement. This raises no particular problems for steady flow, but is not so straightforward when the flow is unsteady. In an unsteady flow, the pressure of the water at a point in the medium varies with time and this may lead to changes in the deformation of the medium itself owing to the weight of the overburden. This is discussed more extensively later.

Similarly, the equations of motion must be satisfied. For the details of the flow through the pore space of the medium, the Navier-Stokes equations are most likely to be relevant. The equations could be extended to cover turbulent flow by taking averages with respect to time. However, turbulent groundwater flow is not a very common phenomenon and we will make only brief reference to it. The Navier-Stokes equation for the x-component of velocity for a Newtonian fluid in iso-thermal motion is

$$\frac{Du}{Dt} = -\frac{1}{\rho}\frac{\partial}{\partial x}(p + \rho gz) + \nu\nabla^2 u$$

and there are similar equations for the other components of velocity. The equation is for unit volume of the fluid, the axis $0z$ is vertical and p is the pressure of the liquid at a point whose elevation is z. The piezometric head h is defined by

$$h = \frac{p}{\gamma} + z$$

and the quantity gz is known as the gravitational potential. Provided the density, ρ, is constant, the gravitational potential is a driving force in the sense that its spatial gradients cause the motion.

In partially saturated soils there is a discrete pressure difference at each liquid-gas interface. This is known as the capillary pressure, p_c, and

$$p_c = \sigma\left(\frac{1}{r_1} + \frac{1}{r_2}\right) = p_{\text{gas}} - p_{\text{liquid}}$$

where σ is the surface tension and r_1 and r_2 are the local major and minor radii of curvature of the interface. Where the gas pressure is atmospheric

$$p_{\text{liquid}} = -p_c$$

and $-p/\gamma(= \Psi)$ is known as the capillary potential. This is treated more fully later.

Flow through a porous medium is affected by the properties of the fluid and those of the medium. The concepts of hydrodynamics apply generally, but there are additional features to be considered. Attention is confined to flow of an incompressible homogeneous fluid, usually water, but it should be borne in mind that such flows comprise only part of a large field of study which includes the flow of stratified liquids, immiscible liquids, mixtures of liquids and gases, etc. In the next three sections the properties of porous media and of water and their interaction are discussed.

1.2 Properties of porous media

A porous medium possesses a network of interconnected voids, passages and fissures through which it is possible for a fluid to penetrate the medium. Usually the medium is assumed to consist of discrete particles, which form voids of varying sizes, and each void or pore is taken to be connected to others by constricted passages. The whole forms a complex interconnected network of irregular passages through which a fluid may flow. A void which is not connected to others will not take part in the flow process. In these tortuous three-dimensional passages the fluid is subject to repeated expansions and contractions, bifurcation and confluence. This is necessarily accompanied by accelerations and decelerations, and dissipation of mechanical energy. The velocity distribution across any pore may resemble that in capillary tube but it is essentially non-uniform in the direction of flow, although the flow may be steady. However, taken macroscopically over an area large compared with the pores, the discharge per unit of area normal to the direction of flow (analogous to velocity) will be much more uniform. In this wider view, details of the flow are lost, but much is gained, for this macroscopic behaviour can be described mathematically. It would be very difficult to get any useful information from a mathematical description of the details of the flow through the effective pore space.

Classification of porous media is another difficult task. They may range from materials through which a fluid would penetrate only under extreme pressure difference to large geologic formations of karstic limestone, where the channels may be of substantial size and far apart. We will exclude caverns and large fissures and classify porous media mainly according to the size of the intergranular spaces.

An important group of porous media are those in which only mechanical forces act on the fluid flowing in the voids or intergranular spaces. These are the forces due to pressure gradients, gravity, inertia and friction. The walls of the voids affect the flow only through viscous friction. Beds of gravel and sand are examples of this type of porous medium.

Parameters which are useful in quantitative description of the pore space in a porous medium are its porosity, its specific internal area and the pore size. The porosity, n, is defined as the ratio of pore volume to total volume and is a measure of the fluid capacity of the medium. One can also speak of the effective porosity if only the inter-connected pore space is considered. The specific internal area a_s is the ratio of the internal area of all the void boundaries to the bulk volume and has dimensions $[L^{-1}]$. It is obviously difficult to measure. Of the various techniques for estimating specific area, such as adsorption, photomicrographs, fluid flow, heat conduction by a gas, etc., the adsorption technique is most reliable. It depends on the relationship between the surface area of a solid and the adsorption of a vapour, or an ionic adsorption.

Definition of pore size is extremely desirable, but unfortunately it appears to be impossible at present. The particle size of the medium is often used to characterize the pore size, ignoring the fact that particles of the same size can be accompanied by a wide range of pore sizes depending on the packing and that definition of the grain size itself is not a simple matter. Various other terms are in use, such as pore size distribution and tortuosity, but these are too specialized to consider here.

Theoretical studies of the packing of spheres are of obvious value in contributing to understanding of porosity and the size and shape of voids. It has been shown that, for uniformly sized spheres of any radius, the tightest packing is rhombohedral and the loosest uniform assemblage is the cubic packing. The properties of these two cases are as follows:

	Cubic	Rhombohedral
Volume of unit cell	8.00 R^3	5.66 R^3
Volume of unit pore	3.81 R^3	1.47 R^3
Porosity	47.67%	25.95%

Hence the porosity of an assemblage of uniformly sized spherical particles can vary from 25.95% upwards. However, the loosest stable packing of spheres that has been described gives a porosity of 87.5%.

In natural soils the grains are not spherical, nor is there only one size or shape present. Observed porosities range through the following values:

Fine to medium gravel (2–20 mm)		30–40%
Sands	2 mm > d > 0.6 mm	39–41%
	0.6 mm > d > 0.2 mm	41–48%
	0.2 mm > d > 0.06 mm	44–49%
Loam		35–50%
Clayey soils		40–55%
Peaty soils		60–80%

For small particles the porosity may range from 50% for particles of about 0.02 mm to 95% for particles smaller than 0.002 mm. In natural sands, with a range of particle size, the porosity can vary between 12% and about 35%. It should also be noted that the geometry and porosity of the medium depends on stresses due to the weight of the overburden (Tiller, 1953, 1955).

If the porous medium is clay, or some material of very small particle size, then, in addition to the mechanical forces, molecular and electro-chemical forces are important. Openings in which adhesive forces between the molecules of the walls and those of the fluid affect the motion of the fluid are known as capillaries. Still smaller spaces are frequently called force spaces. These are small enough for the molecular structure of the liquid to be significant. In clays both the chemistry of the percolating water and the mineralogical structure of the clay are important.

1.3 Properties of water

Water itself is a complex liquid, with a dipole type of molecule made up of two hydrogen atoms and one oxygen atom. Although the molecule as a whole is electrically neutral, the electrical charge (averaged with respect to time) is not uniformly distributed and there is a net positive charge in the vicinity of each hydrogen nucleus and a net negative charge near the oxygen nucleus. Of the oxygen atom's eight electrons, two are held near the nucleus in a complete K shell and the remaining six are in the L shell. Two of the six are shared with the two hydrogen atoms, so that the K shell around each hydrogen nucleus is made complete (two electrons). Similar sharing of the electrons of the hydrogen atoms completes the L shell around the oxygen nucleus (eight electrons). There remain three pairs of electrons which are not shared. The structure of a water molecule is sketched in Fig. 1.2 and it will be seen that the intensity of the electrical charges varies from maximum positive values near the hydrogen nuclei to maximum negative values where the two lone pairs are in orbit around the oxygen nucleus. The net charge distribution in the molecule resembles a tetrahedron with two positive and two negative corners. When two water molecules approach

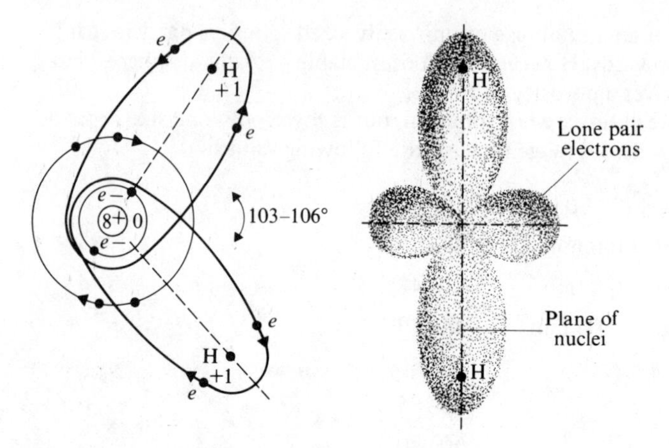

Fig. 1.2. Distribution of average electrical charge in a water molecule.

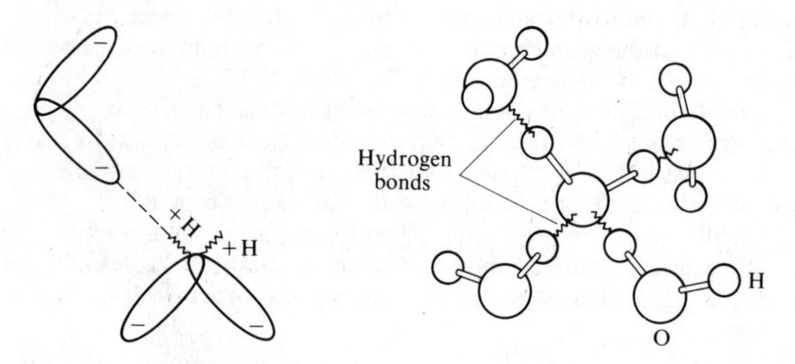

Fig. 1.3. Illustration of hydrogen bond between water molecules.

each other, an electrostatic bond, called the hydrogen bond, can hold the molecules together, Fig. 1.3.

Usually, in water the molecular structure is only partially oriented (unlike other liquids with closely packed structures) and the physical properties are consequently quite different from those of other liquids – high melting point, surface tension, heat of vaporization and dielectric constant, and low coefficient of thermal expansion. When four or more water molecules join together, they form a tetrahedral structure, each molecule being at a vertex of a tetrahedron and joined to three other molecules. One of these three has its hydrogen nucleus bonded to the centre molecule and the other two have their lone pair electrons oriented towards it.

1.4 Interaction of water and the porous medium

These characteristics of the water molecule also affect the bonding of water to solid surfaces. In soils, particularly in clays, water may occur in three locations:

(1) In the crystal lattice of the minerals. This water is strongly bonded to the minerals and is known as high energy water, since temperatures higher than 300° C are required to remove it. Hence, it is of little interest here.
(2) As the inter-layer water between the layers of unit cells in vermiculite and montmorillonite or in tabular openings as in attapulgite, sepiolite and polygorskite. Most of this water can be driven off at 105° C.
(3) Water in the pores of the soil and on the surfaces of the soil grains.

It is the water of group (3) which is of greatest importance here. It has been recognized for a long time that water in close contact with solid surfaces is in a different physical state from the bulk of water: it is more viscous and denser. All solid constituents of soils, irrespective of whether they appear to be chemically inert or not, display electrical properties which can be explained in terms of electrical charging (surface ionization) and dissociation of the surface. The mechanisms for bonding of water to the particles are dealt with in literature on electro-chemistry, and are a complex topic. Only a few brief remarks will be attempted here, which follow the description in *Loose Boundary Hydraulics* by Raudkivi (1976):

'The relevant literature, e.g. Bockris and Reddy (1970), refers to four mechanisms by which a particle may acquire a surface charge. These are

(i) Isomorphous substitution. It takes place when atoms of similar size but of different valencies are somehow swapped within the crystal lattice, e.g. a silicon (valency 4) site is occupied by an aluminium (valency 3) atom and this leads to a negative charge. There is, however, the Pauling's law of isomorphism which states that any deficiency must be balanced in the shortest possible distance from the deficient site. Hence, only local and not surface charge should result.
(ii) Adsorption. This is a process whereby ions from a surrounding electrolyte adhere to a surface and cause it to possess a surface charge. It is not an electrostatic phenomenon whereby just enough cations (positively charged ions) are attracted to a negatively charged surface to balance the charge. The phenomenon seems to depend on the chemical nature of the ions rather than its charge and the adsorbing ion can be oblivious to the charge on the crystal. Anions (negatively charged ions) may contact adsorb on a negatively charged surface. In addition there can be what is known as *super-equivalent adsorption* where the surface adsorbs more ions than is required to balance its charge. As a result the sign of the apparent particle charge changes. Fig. 1.4 shows adsorbed anions (without hydration sheath). Hydrated cations are attracted to this surface and water molecules become aligned. Ideally the charge carried by ions in the outer Helmholtz plane balances the interfacial charge and the potential drop between these two layers is linear. However, usually ions further out in the electrolyte also assist in balancing the surface charge. The resulting neutralizing distribution of ions is referred to as an 'atmosphere' because, as with the atmosphere, the attraction to the interface is counteracted by diffusion. The distribution of potential in this 'diffuse' layer is approximately given by

$$\phi = \phi_\delta \exp[-\kappa(x - \delta)]$$

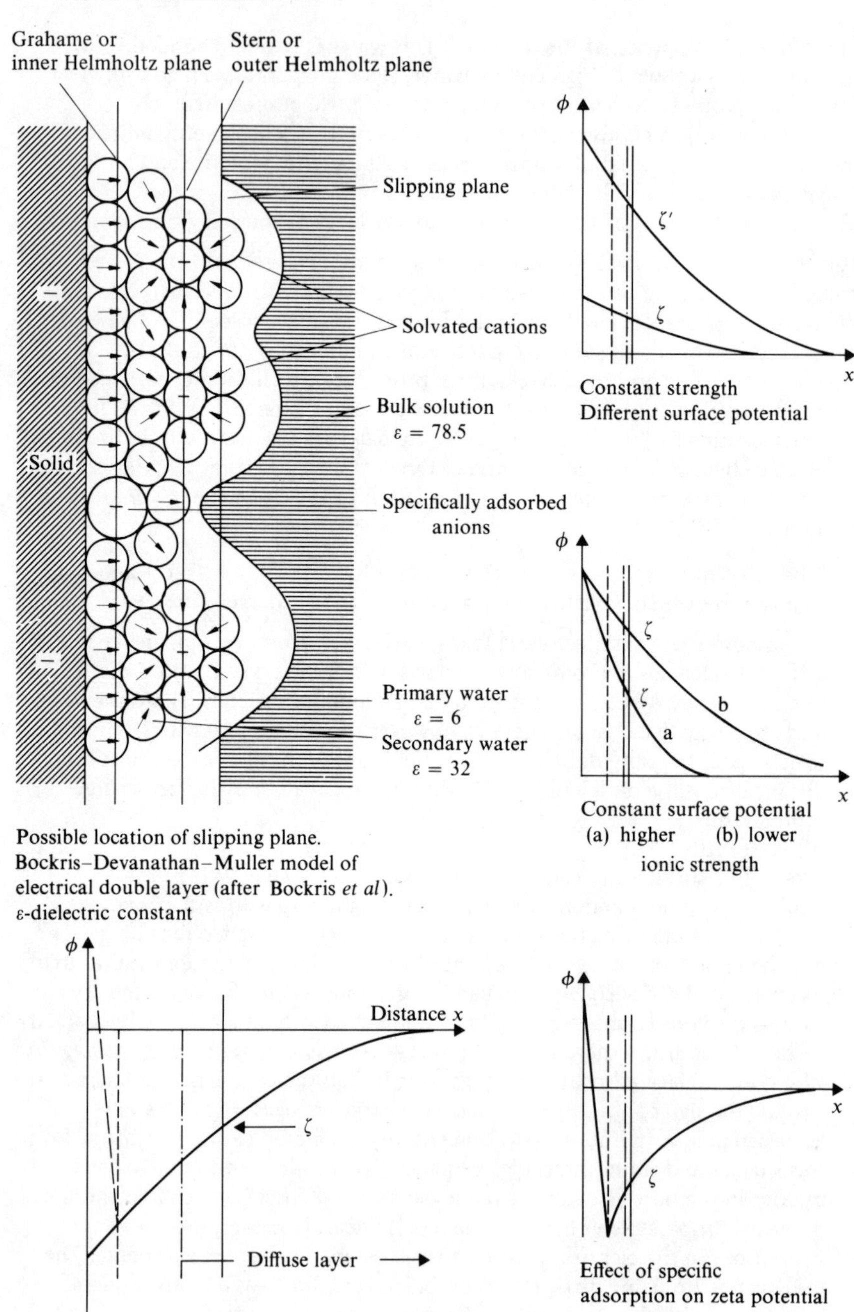

Fig. 1.4. Double layer and Zeta potential, from Raudkivi, 1976.

where

$$\frac{1}{\kappa} = \sqrt{\frac{\epsilon kT}{2e^2LI}} \qquad 1.1$$

is the thickness of the ionic atmosphere, known as the Debye length, and is also a measure of the thickness of the double layer, and ϕ_δ is the potential at the outer Helmholtz plane at $x = \delta$ from the surface. The other symbols are $k = 1.38054 \times 10^{-23}\,J°K^{-1}$ – the Boltzmann constant, T the absolute temperature, ϵ the permittivity of the medium equal to $\epsilon_r\epsilon_0$ (where ϵ_r is the dielectric constant of the material and ϵ_0 the permittivity in a vacuum), $e = 1.60210 \times 10^{-19}C$ the electric charge, $L = 6.02252 \times 10^{23}\,mol^{-1}$ – the Avogadro constant, $I = \frac{1}{2}\Sigma c_i z_i^2$ the ionic strength, c the concentration of the electrolyte (mol/m^3) and z the valency of the counter ion. Substitution of these values leads to

$$\frac{1}{\kappa} = 2.812031 \sqrt{\frac{\epsilon_r T}{\Sigma c_i z_i^2}} \text{ in (pm)}$$

where $\epsilon_r \simeq 80$ for usual temperatures and $p = 10^{-12}$.

When there is relative movement between the charged particle and the electrolyte, slipping will take place between the particle and the electrolyte at some distance out in the diffuse layer, not far from the outer Helmholtz layer (of the order of 10 Å). The potential difference between this plane of shear and the free electrolyte is known as the zeta-potential and can be estimated for a flat surface from

$$\zeta = \frac{v\mu}{\epsilon_r\epsilon_0 H} \qquad 1.2$$

where μ is the viscosity of the electrolyte, v is the velocity relative to fluid in the electric field $H(Vm^{-1})$, and $v/H = v_e$ is the electrophoretic mobility $(m^2 V^{-1} s^{-1})$.

(iii) Ion exchange. Some crystalline substances when mixed with a salt solution will swap ions from the electrolyte with ions from their surface and may thus acquire a net charge. Clay minerals are in general ion exchangers and the more plastic clays have higher exchange capacity. However, not all ion exchangers are cohesive, for example, powdered ion exchange resins.

(iv) Ionization. Clay particle charge is observed to vary not only with concentration of surrounding electrolyte but also with acidity of the electrolyte (pH). The pH at which the particle is uncharged is called the isoelectric point IEP. In a solution which is more acid than the IEP the particle is positive.'

In addition

(1) The surface of many minerals is composed of oxygen, O, and hydroxyl, OH, groups arranged in a hexagonal pattern which could match a similar pattern in hydrogen bonded water, Fig. 1.5. If the crystal lattice of the solid contains excess electrons, owing to isomorphous substitution of cations (positively

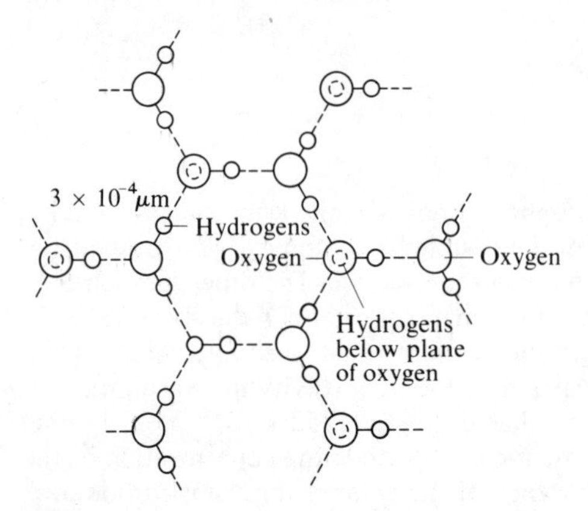

Fig. 1.5. Matching of oxygen and hydroxyl ions in mineral surfaces with hydrogen-bonded water.

charged ions) in the lattice, then covalent (electron sharing) bonds may occur if one of the systems involved is capable of having its lone pair electrons disturbed by the nuclei of the other. The lone pairs of oxygen atoms in the solid surface should be easily disturbed in this way because of the excess of electrons in the lattice. The resulting orientation and bonding can extend through several layers of water molecules.

(2) At low moisture content of the soil, hydration by exchangeable cations is considered to be important.

(3) Orientation of the dipolar water molecules with their axes parallel to the field by the electrical charge on the surfaces changes the nature of water.

(4) The van der Waals forces could also be responsible for water-soil bonding. If the electron atmospheres of the oxygen atoms in the water and the surface of the solid oscillate in phase, the water could be held in a non-directional bond.

Generally, in the vicinity of the surface, the adsorbed water molecules can be closely packed and oriented for many molecular layers and they can be held in a very strong bond. As the liquid water content increases, the film thickness increases. The strength of the bond diminishes rapidly with the distance from the solid surface and the outermost layers of the bonded water can be removed readily by evaporation and transpiration. The bonding force for the innermost layers of water molecules may be as high as – 100 MPa. In contrast the suction of plant roots at the wilting point is of the order of 700–800 kPa. The bonding force, or pore pressure deficiency, may arise from capillary suction, when water is held by capillary action; from the bonding energy when only adsorbed water is present; or from osmotic pressure due to dissolved salts. The latter form is known as solute suction and the former as capillary potential or matrix suction, designated by Ψ, h or τ. When the liquid water content increases so that all the adsorbed water demand is satisfied, the films around the particles can merge to form droplets of water between adjacent grains. This is known as capillary water. Such a liquid-air interface is the same as the meniscus in a capillary tube.

Capillary forces depend on the angle of contact between the media in contact. The latter is a measure of the relative strength of the attractions of the liquid

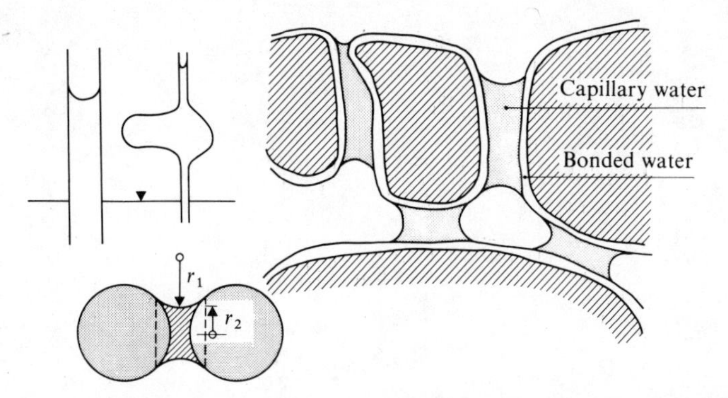

Fig. 1.6. Illustration of bonded water, capillary water and capillary rise.

molecules to each other and to the molecules of the solid. A single capillary surface is balanced either by a similar opposing surface (as in a bubble or a drop) or, when connected with the water table, by a column of water. Note that the shape of the column under the meniscus does not influence the height to which the water rises, Fig. 1.6.

The capillary rise is usually given by semi-empirical relationships. For example, for a capillary tube

$$\frac{p_c}{\gamma} = H = \Psi = (2\sigma/\gamma r) \cos \alpha$$

where σ is the surface tension and equals 7.27×10^{-2} Nm^{-1} for water at 20° C, α is the acute angle at the liquid-solid contact and r is the radius of the tube.

For a capillary between two grains the pressure difference at the curved interface is

$$p_c = \sigma \left(\frac{1}{r_1} + \frac{1}{r_2} \right) = p_{\text{gas}} - p_{\text{liq}}$$

where r_1 and r_2 are the local major and minor radii of curvature of the interface. When $r_1 = r_2 = r/\cos \alpha$, we obtain the above relationship for a capillary tube of radius r.

Leverett (1941) showed that experimental results for capillary rise in unconsolidated sands reduced to a single curve when plotted in the form $J(s)$ versus s where $J(s) = (p_c/\sigma)\sqrt{k/n}$, k is the permeability, n is the porosity, and s is the degree of saturation, equal to the ratio of volume of water to volume of voids. Fig. 1.7 shows typical curves. Brooks and Corey (1964, 1966) showed that p_c/γ, plotted against s_e on log-log scales, yielded straight lines, except as s_e approached one – see Fig. 1.8. Here $s_e = (s - s_0)/(1 - s_0)$ and s_0 is the residual wetting saturation which cannot be removed by hydraulic forces; it is the asymptotic value of s which is approached as p_c increases. The results were described by

$$\ln s_e = -\lambda \ln p_c + \lambda \ln p_b ; p_c \geq p_b$$

where p_b is the bubbling pressure.

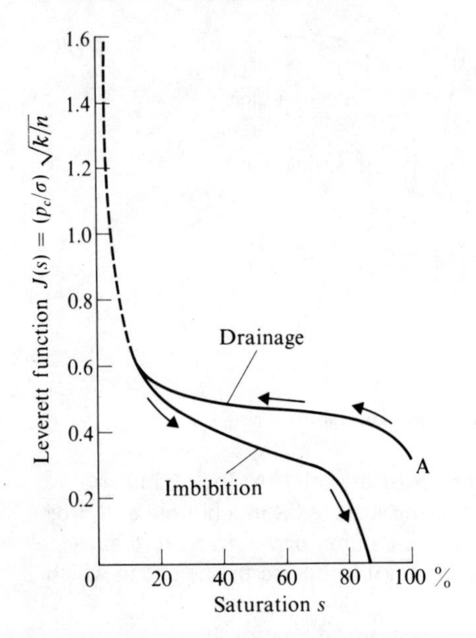

Fig. 1.7. Typical Leverett functions for sand, Leverett, 1941.

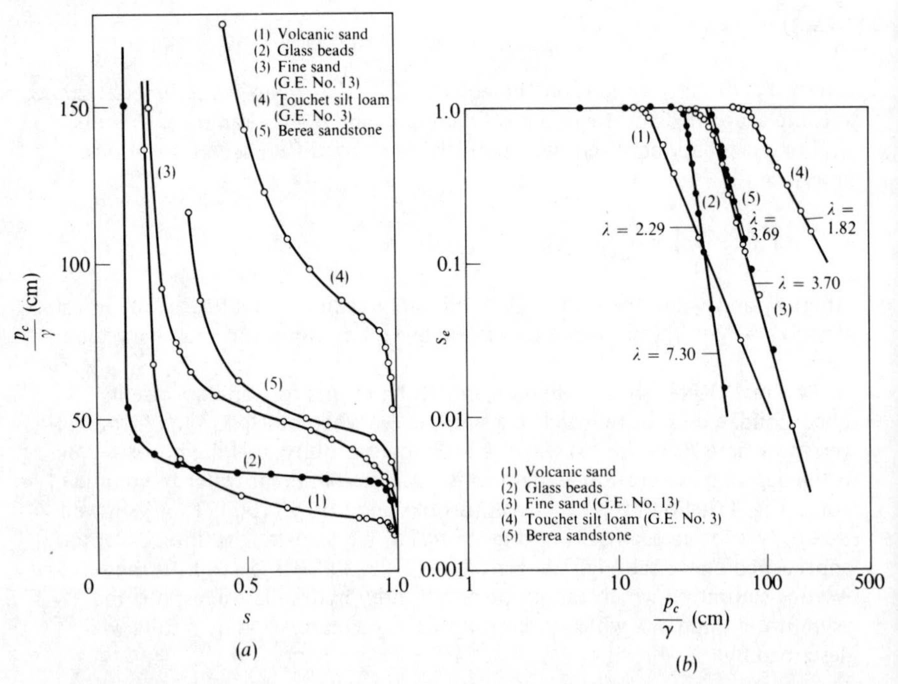

Fig. 1.8. (*a*) Capillary pressure head as a function of saturation for porous materials of varying pore-size distributions. (*b*) Effective saturation as a function of capillary pressure head for porous materials of varying pore-size distributions, Brooks and Corey, 1964.

A number of definitions of the moisture content, in addition to the degree of saturation, are in use. A common one is the moisture content defined as the mass of water per unit bulk volume, i.e. g/cm^3. But this can also be found expressed as the ratio of mass of water to mass of solids in which the water is stored, or as the volumetric moisture content θ which is equal to the ratio of volume of water to total volume, i.e. $\theta = ns$.

Written in thermodynamic form the soil water potential is

$$dP = \left(\frac{\partial P}{\partial p}\right)_{n_w} dp_{n_j} + \sum_j \frac{\partial P}{\partial n_j} dn_j + \frac{\partial P}{\partial n_w} dn_w + \frac{\partial P}{\partial T} dT$$

where P is the chemical potential of water in soil, n_w is mole fraction of water, n_j is mole fraction of solute, p is pressure and T is temperature. The terms on the right hand side account for the effect of hydraulic pressure, osmotic potential, capillary potential and the effect of temperature differences, respectively.

The time required to complete the capillary rise is theoretically infinite. Since the surface tension is not resisted at the beginning, and is balanced by the weight of the water suspended below the meniscus at the end, there is an asymptotic approach to the equilibrium configuration.

The contact angle between dry powdery soils and water is larger than 90° – there is a repulsion between the water and the soil surface. This fact is important in controlling the initial rate of rainfall penetration by resisting the spreading of water over the solid surface.

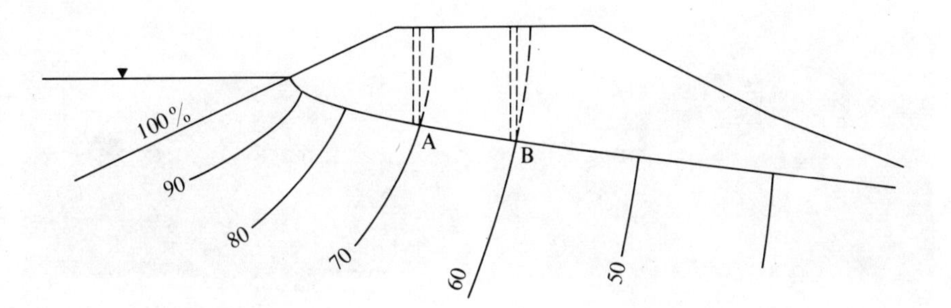

Fig. 1.9. Flow in the capillary fringe.

Provided the water table is not horizontal, water will flow in the capillary fringe in a direction approximately parallel to the water table. Consider a stopbank as illustrated in Fig. 1.9. Assume soil capillaries at A and B rising up from the phreatic line (a streamline along which the pressure is atmospheric) as defined for the case where there is no capillary rise, and that in these after a sufficient period of time the water has risen to an equilibrium height. The water column in the tube A is balanced by the meniscus and the piezometric head (potential value) throughout its height will be 70% approximately and similarly at B. (In general the equipotential is not exactly vertical.) Hence, there is a general potential gradient in the fringe, see Fig. 1.10. However, since the fringe is not fully saturated, the porosity and the permeability are both less than they are below the phreatic line and the flow rate is correspondingly reduced.

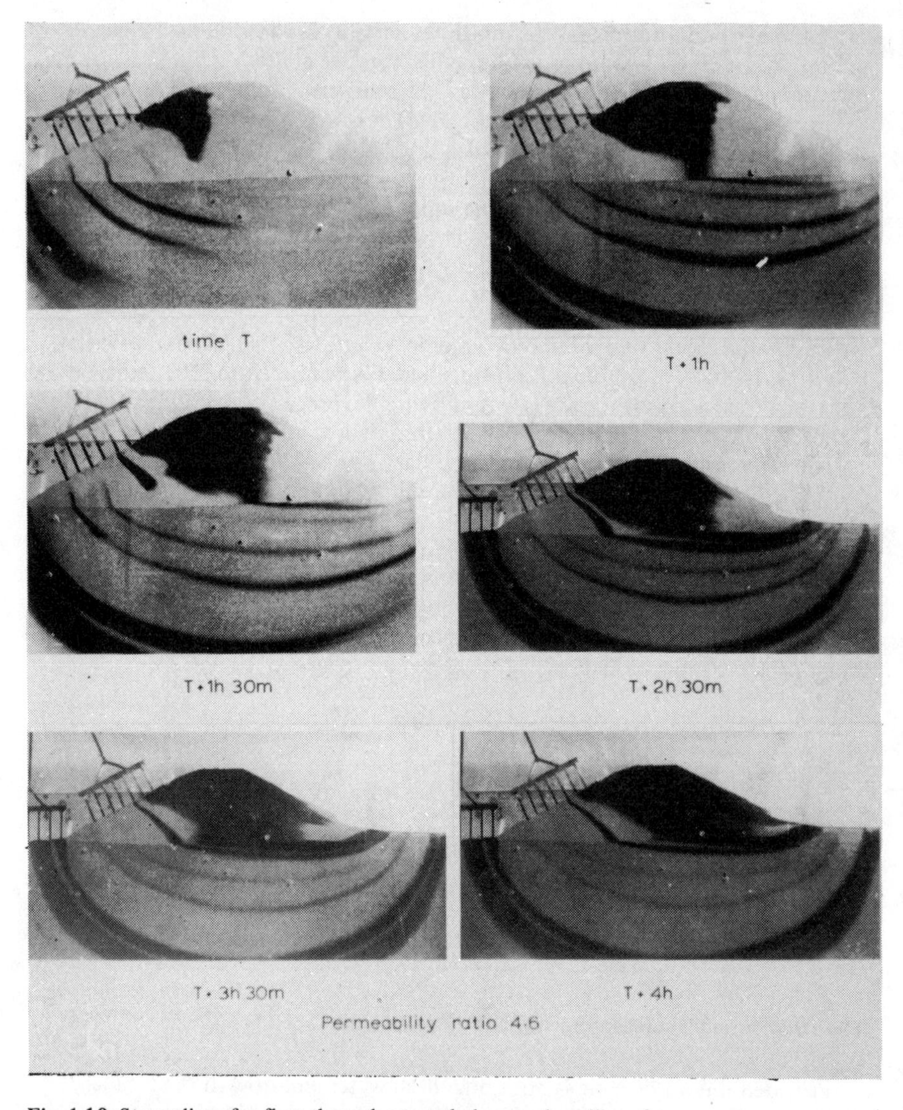

Fig. 1.10. Streamlines for flow through an earth dam, and capillary fringe.

Finally it is important to emphasize that in fine grained soils the chemistry of water becomes important. The permeability of a sandy-clay soil can be very different for waters with different cation or anion (negatively charged ion) content, since the thickness of the bonded film affects the pore size.

Attention is also drawn to the fact that when the fluid flows in a passage of the same order of size as the mean free path of the fluid molecules, the fluid no longer behaves as a continuum. It has been observed with gas flows in such passages that there is slip at the boundaries and the rate of flow is greater than would be expected from theory, Klose (1931), Arnell (1946). This is now referred to as the slip phenomenon, or Knudsen (1950) flow, or Klinkenberg effect (1941).

1.5 Darcy's Law

In 1865, Henri Darcy published in an appendix to his book *Les Fontaines Publiques de la Ville de Dijon* the results of his experiments on the flow of water through granular material. Using a cylindrical sample, the direction of flow being along the cylinder, he found the discharge per unit area of cross-section to be proportional to the gradient of piezometric head in the direction of flow, i.e.

$$\frac{Q}{A} = K\frac{\Delta h}{\Delta l} \qquad 1.3$$

Here, Q is the discharge through the sample, A is the gross area of the sample's cross-section, Δh is the head lost in a length Δl and K is constant for a given sample. This expression, and various rearrangements of it, have been named Darcy's law and it is the basic relationship in quantitative study of the flow of fluids through porous media.

The constant K, which has the dimensions [L/T] was called the transmission constant by Slichter and others. It is now known as the hydraulic conductivity or the coefficient of permeability, an unfortunate name, since K depends not only on the permeability of the soil (properly called) but also on the properties of the fluid. This must be the case, since the rate of flow will depend on the fluid and no fluid properties appear explicitly in equation 1.3. It follows that variations in fluid properties will cause changes in K, i.e.

$$K = f(\mu, \gamma, d) \qquad 1.4$$

where μ is the coefficient of viscosity and is a measure of the resisting force, γ is the specific weight, a measure of the driving force and d is the particle size, assumed to be a measure of the pore size. Thus, if the mean grain size is increased, but the shape and size distribution of the grains remain unchanged, the resistance of flow will decrease and K will increase because of the bigger voids.

Dimensional analysis applied to equation 1.4 yields

$$K = c\frac{\gamma}{\mu}d^2 = k\frac{\gamma}{\mu} \qquad 1.5$$

in which the fluid properties are kept separate from the relevant soil property. The product $cd^2 = k$ depends on the characteristics of the porous medium only and is thus correctly called the coefficient of permeability or the permeability. The factor c has to account for the characteristics of the medium, i.e. its porosity, the range and distribution of grain size, the shape of the grains and their arrangement, although electro-chemical effects are omitted. The dimensions of k are $[L^2]$, and in petroleum engineering, k is measured in units called darcy, defined as follows: 'a porous medium is said to have a permeability of one darcy if a single-phase fluid of one centipoise viscosity that completely fills the pore space of the medium will flow through it at a rate of 1 cm^3/s per cm^2 of cross-sectional area under a pressure gradient of 1 atm. per cm.' (Bear, 1972).

From equation 1.5.

$$k = \frac{\mu Q/A}{\mathrm{d}p/\mathrm{d}l}$$

and k will be 1 darcy if μ is 1 centipoise or 10^{-2} dyne s/cm^2, Q/A is 1 cm^3/s per cm^2 and $\mathrm{d}p/\mathrm{d}l$ is 1 atmosphere per cm or 1.0132×10^6 dyne/cm^2. Hence,

$$1 \text{ darcy} = 0.987 \times 10^{-8} \text{ cm}^2$$

For water at 20°C, a medium of permeability 1 darcy would have a hydraulic conductivity K of 9.613×10^{-4} cm/s.

Many empirical and theoretical formulae for k in terms of the properties of the matrix of the porous medium have been proposed. For example, there is the empirical relationship due to Krumbein and Monk (1942)

$$k = 0.617 \times 10^{-11}\, d^2$$

where k is the permeability in cm^2 and d is the grain size in μm. Another is

$$k = \frac{1}{m}\left[\frac{(1-n)^2}{n^3}\left(\frac{\alpha}{100}\,\Sigma\,\frac{P}{d_g}\right)^2\right]^{-1}$$

This is based on dimensional reasoning by Fair and Hatch (1933). Here m is a packing factor ($\sim$5 by experiment), α is a sand shape factor which ranges from 6.0 for spherical grains to 7.7 for angular grains, n is the porosity of the soil, P is the percentage of grains passing one sieve and held on the next, and d_g is the geometric mean of these sieve sizes.

Flow through a porous medium can be thought of as flow around small spheres at low velocity, a phenomenon studied by Stokes. Neglecting the inertia terms in the Navier-Stokes equations, Stokes obtained for the resistance force $F = 3\pi\mu U d$, which is confirmed by experiment for $R_e < 1$. Rearrangement of this equation, with the volume of the sphere as a unit of volume yields

$$\text{Force per unit volume} = \frac{18\mu U}{d^2}.$$

Now Darcy's law may be rearranged as follows:

$$\text{Force per unit volume} = \frac{\mathrm{d}p}{\mathrm{d}l} = \gamma\,\frac{\mathrm{d}h}{\mathrm{d}l}$$

$$= \frac{\gamma Q/A}{K} = \frac{1}{c}\,\frac{\mu Q/A}{d^2}$$

This suggests that Darcy's linear relationship also applies over the range where inertia effects are negligible. And indeed Darcy's law can be deduced from the Navier-Stokes equations (Hubbert, 1956). The derivation applies statistical considerations in the process of simplifying the complicated details of the flow. A simplified version of analysis is given by De Wiest (1965).

For a more general analysis, that is, without the restriction that inertia forces are negligible we can expect that

$$\frac{\mathrm{d}p}{\mathrm{d}l} = f(\mu, U, \rho, d)$$

ignoring the effects of particle shape and distribution of size, and electrochemical forces.

Dimensional analysis yields

$$\frac{\mathrm{d}p}{\mathrm{d}l} = NU^n d^{n-3} \mu^{2-n} \rho^{n-1} \qquad 1.6$$

where N is a number.

When $n = 1$, equation 1.6 reduces to Darcy's law

$$\frac{\mathrm{d}p}{\mathrm{d}l} = \frac{N\mu U}{d^2} \qquad 1.7$$

or

$$U = \frac{Q}{A} = \frac{1}{N}\,\frac{\gamma d^2}{\mu}\,\frac{\mathrm{d}h}{\mathrm{d}l} \qquad 1.8$$

and the number N is identified with c in $k = cd^2$, i.e. $N = 1/c$. Thus, N depends solely on the shape and size distribution of the particles.

In the range of Reynolds number where inertia forces predominate, $n = 2$ and equation 1.6 yields

$$\frac{\mathrm{d}p}{\mathrm{d}l} = 2N\frac{1}{d}\,\frac{\rho U^2}{2} \quad \text{or} \quad \frac{\mathrm{d}h}{\mathrm{d}l} = 2N\frac{1}{d}\,\frac{U^2}{2g} \qquad 1.9$$

Equation 1.9 will be recognized as the Darcy-Weisbach formula familiar in the flow of fluids in pipes if f is substituted for $2N$. In fact, equation 1.6, compared with the Darcy-Weisbach formula gives

$$f = \frac{2N}{Re^{2-n}} \qquad 1.10$$

where f is the Darcy-Weisbach friction factor and N is assumed to take into account the relationship between grain size and pore size.

Experiment shows that Darcy's law is valid for Reynolds number less than 1. However, the law does not depart seriously from the observed results for Re up to 10. The inertial range is fully established at $Re > 1000$ and equation 1.9 is a good approximation for Re as low as 600. Fig. 1.11 shows the Fanning friction factor (equal to a quarter of the Darcy-Weisbach f) as a function of Reynolds number, (Rose (1945)).

We emphasize again that U is the bulk velocity, that is, discharge per unit of area of the porous medium. In the voids, where the flow occurs, the actual velocity is greater. Bakhmeteff and Feodoroff (1937) express the velocity in the pores U_p in terms of the bulk velocity U as

$$U_p = \frac{U}{n^{2/3}}$$

The factor $1/n^{2/3}$ is justified by the argument that in an elementary cube of bulk volume V_1, the volume of the pores is $V_2 = nV_1$ and the ratio of the areas of corresponding faces of the 'cube of voids' is $1/n^{2/3}$, Fig. 1.12. Thus, $V_1 = s_1^3$, $A_1 = s_1^2 = V_1^{2/3}$; $V_2 = nV_1, A_2 = (nV_1)^{2/3}; A_1/A_2 = 1/n^{2/3}$ and $s_1/s_2 = 1/n^{1/3}$. Similarly, the void size is taken to be $d_p = dn^{1/3}$. The Reynolds number for the flow in the pores is now given by

$$Re_p = (U/n^{2/3})\, dn^{1/3}/\nu = Re/n^{1/3}$$

and the friction factor f (equation 1.9) by $f_p = fn^{5/3}$. When these coordinates are used, experimental data plot more as a single function.

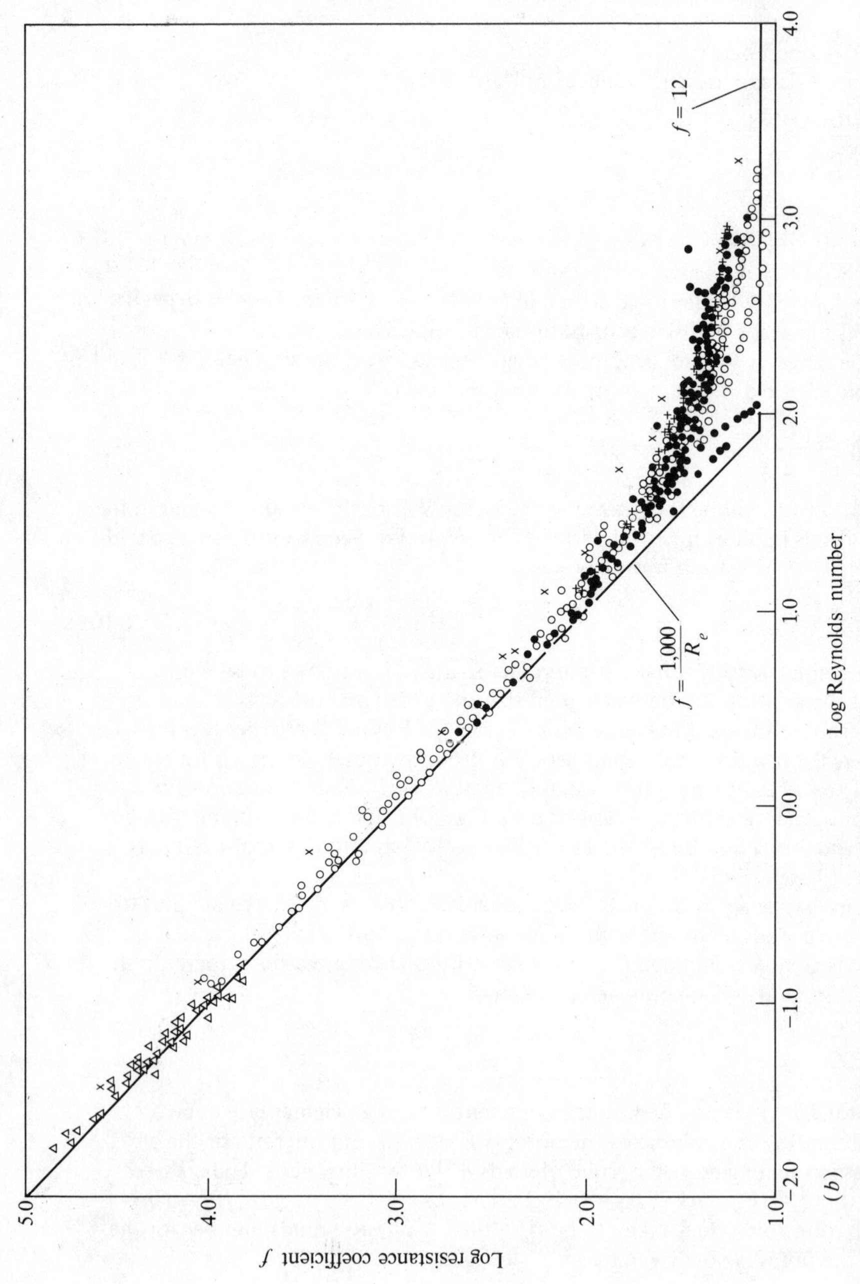

Fig. 1.11. Resistance coefficient plotted against Reynolds number for a granular bed, according to Rose, 1945. Correlated from the data of • Burke and Plummer, + Saunders and Ford, △ Mavis and Wilsey, × Rose, ○ Bakhmeteff and Feodoroff. f is Fanning's friction factor equal to a quarter of the Darcy-Weisbach f.

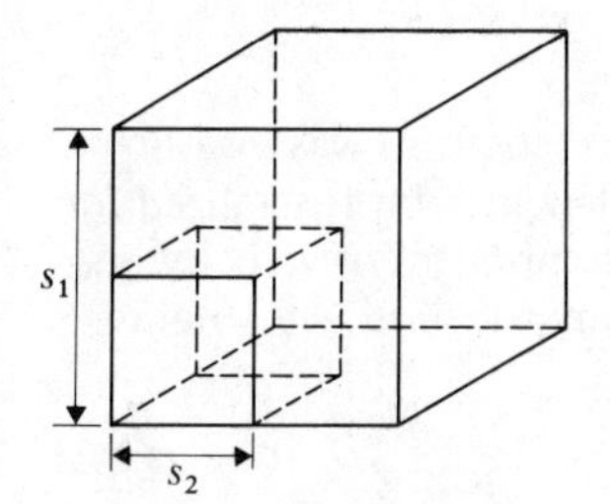

Fig. 1.12. Bulk volume and volume of voids.

When all the variables in the equation $K = k^{\gamma/\mu}$ are functions of space, problems of definition as well as analytical difficulties arise. Flow through an anisotropic medium will be discussed in a later section. Here, it should be noted that even an isotropic material will show a variation in k with the stresses due to the weight of the overburden. The empirical equation

$$k = A(P - p)^{-m}$$

is a good approximation, where P is a scalar overburden pressure and p is the fluid pressure.

Some typical values of Coefficient of Permeability

Soil type	K cm/s.
Clean gravel	≥ 1.0
Clean coarse sand	1.0 –0.01
Sand mixture, clayey sand	0.01 –0.005
Fine sand	0.05 –0.001
Sandy loam	0.005 –0.003
Silty sand	0.002 –0.0001
Peat, little decomposed	0.006 –0.002
Peat, moderately decomposed	0.0008–0.0002
Silt	0.0005–0.00001
Clay	< 0.000001

Table 1.1 by Irmay from Bear *et al.* (1968) gives K and k in terms of $\log_{10}$ as used by the US Bureau of Reclamation (md ≡ milli-darcy).

Table 1.1. Typical values of hydraulic conductivity and permeability.

$-\log_{10} \cdot K$(cm/s): −2, −1, 0, 1, 2, 3, 4, 5, 6, 7, 8, 9, 10, 11

<table>
<tr><td>Permeability</td><td colspan="4">Pervious</td><td colspan="4">Semipervious</td><td colspan="5">Impervious</td></tr>
<tr><td>Aquifer</td><td colspan="5">Good</td><td colspan="4">Poor</td><td colspan="4">None</td></tr>
<tr><td rowspan="2">Soils</td><td colspan="2">Clean gravel</td><td colspan="3">Clean sand or sand and gravel</td><td colspan="4">Very fine sand, silt, loess, loam, solonetz</td><td colspan="4"></td></tr>
<tr><td colspan="4"></td><td colspan="2">Peat</td><td colspan="3">Stratified clay</td><td colspan="4">Unweathered clay</td></tr>
<tr><td>Rocks</td><td colspan="4"></td><td colspan="3">Oil rocks</td><td colspan="2">Sandstone</td><td colspan="2">Good limestone, dolomite</td><td colspan="2">Breccia, granite</td></tr>
</table>

$-\log_{10} \cdot k$(cm²): 3, 4, 5, 6, 7, 8, 9, 10, 11, 12, 13, 14, 15, 16

$\log_{10} k$(md): 8, 7, 6, 5, 4, 3, 2, 1, 0, −1, −2, −3, −4, −5

1.6 Potential flow analogy

Darcy's law was derived from experiments in which the length Δl was measured in the direction of flow. If it is now assumed that the law may be generalized for application to a three-dimensional flow, so that each component of velocity is proportional to an appropriate rate of change of piezometric head, we obtain

$$u = -K\frac{\partial h}{\partial x}$$

$$v = -K\frac{\partial h}{\partial y}$$

$$w = -K\frac{\partial h}{\partial z}$$

or, in vector notation

$$\mathbf{q} = -K \operatorname{grad} h \qquad 1.11$$

Here, $\mathbf{q}$ is the bulk velocity and a minus sign has been introduced to accommodate the fact that the direction of flow is in the direction of decreasing piezometric head h. If we now define the quantity ϕ by

$$\phi = -Kh + \text{constant} \qquad 1.12$$

and K does not vary through the porous medium, then

$$\mathbf{q} = -\operatorname{grad} \phi \qquad 1.13$$

The important fact emerges that there is an analogy between flow in a porous medium and potential flow, the quantity (Kh + constant) bearing the same relation to the bulk velocity in the one as the velocity potential bears to velocity in the other. The analogy is of great practical importance in some two-dimensional flows, for the flow nets of certain two-dimensional potential flows can be adapted readily to the solution of problem in flow through porous media. It cannot be overemphasized that this application of the theory of potential flow is valid only because an analogy exists. It is not implied that the fluid flow through the tortuous intergranular passages is irrotational – it is not, and a velocity potential in the literal sense does not exist. Practical application of a similar analogy is also possible in a case where $\phi = \frac{1}{2}Kh^2$ + constant, as will be shown later. The basic elements of elementary potential flow patterns are listed in Appendix I. For additional treatment the reader is referred to Raudkivi and Callander (1975).

1.7 Non-Darcian flows

Darcy's law is not always valid, even for laminar flows. Many heavy soils and clays show non-Darcian behaviour at low Reynolds and in Fig. 1.13 there is a selection of experimental curves showing how the bulk velocity q varies with the gradient of piezometric head dh/dx. Darcy's law appears in these coordinates as a straight line through the origin.

No doubt some of the reported departures from Darcy's law are caused by

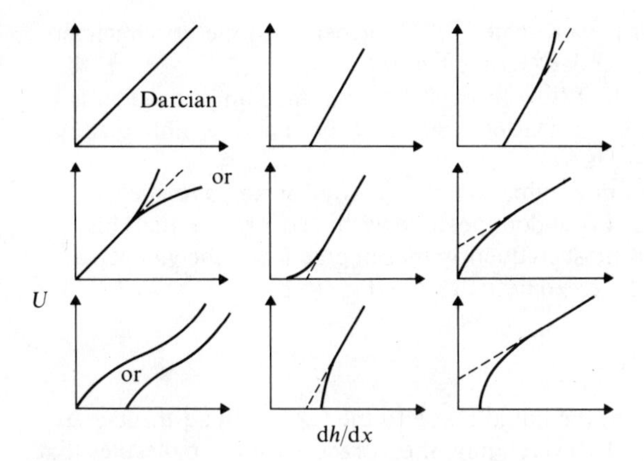

Fig. 1.13. Types of curves for bulk velocity as a function of gradient of piezometric head, after Kutilek, 1972.

experimental error, arising from such things as swelling of the soil, entrapped air and bacterial growths, but there are several reasonable explanations for disagreement with the linear law. We have already referred to the ordered arrangement of water molecules in clays and this quasi-crystalline structure is a reason for non-Darcian behaviour, as discussed by Swartzendruber (1962, 1968). There is the effect of the streaming potential. As water moves near a clay surface it carries along some of the cations in the diffuse layer. The cations are attracted electrically to the clay particles and the attraction opposes the motion of the cations. This in turn causes a drag on the water moving through the soil. The potential difference due to this migration of cations is the streaming potential and it acts in the opposite direction to the hydraulic potential (Kutilek, 1969, Kirkham and Powers, 1972). Changes in the geometrical arrangement of the particles of the sample may be another cause of non-Darcian behaviour. For example, small particles may be transported from one place and deposited in another, leading to the opening of some pores and blocking of others. Non-Newtonian behaviour of water in capillary spaces could also be relevant as suggested by Irmay in Chapter 5 of Bear *et al.* (1968). A group of research workers in Russia has suggested that water in capillaries has the properties of a Bingham body. Slip in very small pore spaces has already been referred to.

1.8 Flow at large Reynolds numbers

As the pore size or the hydraulic gradient or both increase, the linear relationship between the flow rate and gradient ceases to be valid. For this region Forchheimer (1901) proposed a relationship of the form

$$\frac{dh}{dl} = S_e = a\,|\,q\,|+b\,|\,q\,|^2$$

where a and b are constants. He complemented this by adding a further term $c\,|\,q\,|^3$ and later by a relationship where the exponent 2 was made variable between 1.6 and 2.0. Polubarinova-Kochina (1962) generalized the Forchheimer equation and added a time dependent term $c\partial |q| / \partial t$.

For detailed study of turbulent flow through a porous medium the reader is referred to Bear (1972), Scheidegger (1960), Sunada (1963) and its bibliography, Kirkham (1967) and Bachmat (1967).

The practical difficulties of determination of so many constants for field application are enormous and it is seldom possible within cost restraints. This non-linear flow relationship is most frequently met in practice at the vicinity of a well screen or filter where the piezometric gradient is steep.

1.9 Partially saturated soils

A water molecule escaping from the liquid phase to the vapour phase through a horizontal free surface does so by overcoming the attraction of the molecules that surround it. In a partially saturated soil, the forces to be overcome are increased because of the curvature of the interface between liquid and gas and the proximity of solid surfaces. The vapour pressure in the atmosphere in the voids is always less than the saturation vapour pressure for air at the same temperature over a flat surface. Typical curves showing the variation of vapour pressure and relative humidity with moisture content are shown in Fig. 1.14. These porous media show a sharp change in this characteristic at a relative humidity of about 0.96. Below that value a small change in water content results in a large change in relative humidity. Above it, large changes in water content effect only small changes in relative humidity. The moisture content at which this change occurs is sometimes referred to as the wilting point and it is an indication of a boundary between predominance of adsorptive and capillary forces.

Assuming that the water vapour behaves like a perfect gas, the vapour density, ρ, in the pore space can be related to the relative humidity H_r by

$$\rho = \rho_s H_r$$

where ρ_s is the saturation vapour density at the same temperature. Edlefsen and

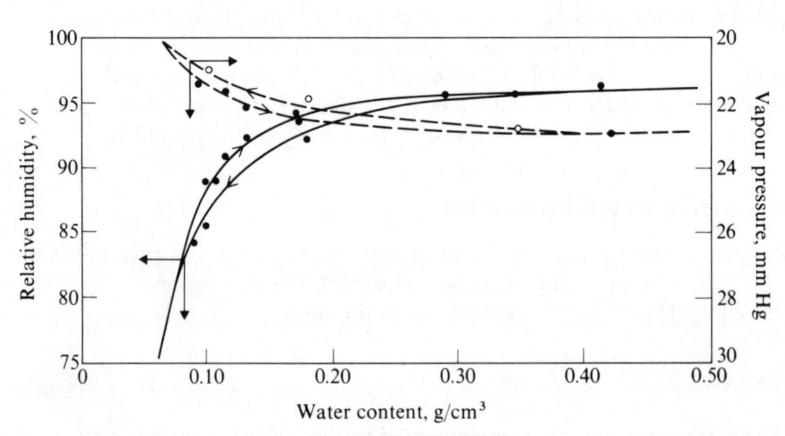

Fig. 1.14. Relationship between relative humidity and moisture content in greywacke loam, Nguyen, 1974.

Anderson (1943) showed that, above the wilting point in the capillary condensation range of soil moisture, H_r is related to matrix suction head ψ by

$$H_r = \exp\,(g\psi/RT)$$

where ψ is proportional to the surface tension σ, R is the gas constant for water vapour (461.37 J kg^{-1} K^{-1}), T is the absolute temperature in Kelvin and $g\psi$ is equivalent to the total specific energy or Gibbs' free energy Δf_s of the soil moisture. The corresponding differential equation is

$$\frac{\partial H_r}{\partial T} = \left[\frac{1}{\sigma}\frac{\partial \sigma}{\partial T} - \frac{1}{T}\right] H_r \ln H_r$$

In the normal range of temperature $(1/\sigma)/(\partial\sigma/\partial T) \sim 2 \times 10^{-3}\text{K}^{-1}$ so that, with $H_r = 0.98$ the value of $\partial H_r/\partial T$ is of the order of $3 \times 10^{-5}\text{K}^{-1}$; clearly the relative humidity is not sensitive to temperature in this range.

Below the wilting point there is no capillary water, the liquid water being held in place by adsorption. As much as 350 kJ kg^{-1} more energy may be required to evaporate adsorbed water from a film a few molecular layers thick than from a free water surface. Equilibrium of water vapour with adsorbed water is described by

$$\frac{\partial(\ln e)}{\partial T} = \frac{L_e{*}}{RT^2}$$

where L_e* is the latent heat of evaporation of the adsorbed water and e is the partial vapour pressure. Over a free water surface the corresponding relationship is

$$\frac{\partial(\ln e_s)}{\partial T} = \frac{L_e}{RT^2}$$

Hence, by subtraction

$$\frac{\partial(\ln e/e_s)}{\partial T} = \frac{L_e{*} - L_e}{RT^2}$$

and since $H_r = e/e_s$

$$\frac{\partial H_r}{\partial T} = H_r\,\frac{L_e^* - L_e}{RT^2}$$

Using Orchiston's data (1953), $L_e* - L_e = 350$ kJ kg^{-1}, with $H_r = 0.60$, $\partial H_r/\partial T \sim 4 \times 10^{-3}K^{-1}$. Clearly temperature has little effect on relative humidity in this range also.

The fact that water occurs in both liquid and vapour phases in partially saturated soils means that its movement occurs in different modes depending on the level of saturation. When the moisture content is low the transfer is in the form of molecular diffusion rather than viscous flow. The migration of adsorbed liquid has been shown to be more than an order of magnitude less than diffusion of the vapour. Since the humidity does not depend on temperature, we can restrict our discussion of vapour diffusion to isothermal conditions. This is not to say that a temperature gradient has no effect on migration of water vapour.

First, the flow of liquid water is described by a diffusion equation. In the capillary zone the water flows under the action of gradients of the capillary potential. For example, the flow of liquid water in the vertical direction is given by Darcy's law as

$$q_l = -K(\theta)\frac{\partial\psi}{\partial z}$$

where the permeability of the unsaturated soil is K and depends on the volumetric moisture content θ. The effect of gravity on the flow has been ignored. The continuity equation ensures that

$$\frac{\partial q_l}{\partial z} = \frac{\partial\theta}{\partial t}$$

so that

$$\frac{\partial\theta}{\partial t} = \frac{\partial}{\partial z}\left\{K(\theta)\frac{\partial\psi}{\partial z}\right\}$$

With the diffusivity for the liquid phase defined by

$$D_l(\theta) = K(\theta)\frac{\partial\psi}{\partial\theta}$$

this becomes

$$\frac{\partial\theta}{\partial t} = \frac{\partial}{\partial z}\left\{D_l(\theta)\frac{\partial\theta}{\partial z}\right\} \qquad 1.14$$

which is the same as the diffusion equation. The diffusivity is not a single valued function of the moisture content. It has a hysteresis which depends on the history of wetting and drying of the soil.

This equation can be extended to describe flow in three-dimensional space. Also, flow of water vapour and adsorbed water can be described similarly with concentration dependent diffusivities $D_v(\theta)$ and $D_a(\theta)$. Jackson (1964) and Rose (1969) added all these diffusion components to yield one diffusivity which describes isothermal transport of both phases, as suggested earlier by Philip (1956):

$$D(\theta) = D_l(\theta) + D_v(\theta) + D_a(\theta)$$

The component $D_l(\theta)$ has hydrodynamic or momentum exchange characteristics while the other two are diffusivities in the classical sense depending on random molecular movement. Fig. 1.15 shows a typical relationship for $D(\theta)$ as plotted by Philip (1957); see also Jackson (1964).

When there is a temperature gradient, the Philip-de Vries model for one dimensional moisture transport is described by the question

$$q = -D(\theta)\frac{\partial\theta}{\partial x} - D(T)\frac{\partial T}{\partial x}$$

where $D(T)$ is the thermal diffusivity. This shows that transport of water is in the direction of decreasing temperature and if condensation occurs at the cold end, a

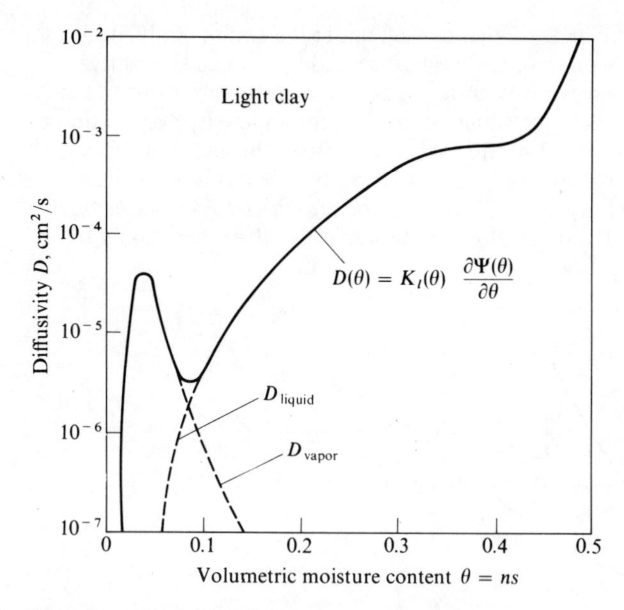

Fig. 1.15. Diffusivities D_{liq}, D_{vap} and moisture diffusivity $D(\theta)$ as a function of volumetric moisture content θ, plotted by Philip, 1957, after data of Moore, 1939, for Yolo light clay ($\theta > 0.22$). Note that for $\theta < 0.07$, $D(\theta)$ is indistinguishable from D_{vap}; and that for $\theta > 0.10$, $D(\theta)$ is indistinguishable from D_{liq}.

steady flow can occur. The phenomenon is important in relation to accumulation of water under road pavements. The rate of moisture transfer depends on the soil type, its dry density, the degree of saturation and the temperature gradient. Sands and silty loams transfer moisture more freely than clays do, less dense soils more freely than dense soils. The rate of flow is greatest when the moisture content is between the plastic limit and a moisture content slightly less than the field capacity (Raudkivi and Nguyen, 1975).

Moisture moves under the action of temperature gradient in both liquid and vapour phases. Transport of the vapour is described by molecular diffusion. For the liquid phase, Darcy's law for a partially saturated medium is relevant. Evaporation and condensation may occur locally. For example, liquid water may be held between soil particles by capillary action, with a meniscus on either side. In a temperature gradient vapour can condense on one side and liquid evaporate on the other. There would be a flux of water through the liquid held by the grains.

It should be noted that temperature gradients play a very different role in the flow of water in geothermal fields. For a discussion of heat and mass transfer in these circumstances Bear (1972) should be consulted.

1.10 Infiltration

When snow or rain falls on the ground, some water flows over the surface and some soaks into the ground. Infiltration is the name given to the process whereby water flows into and through the soil under the action of gravitational and capillary forces. It is relevant to design of irrigation works and drainage works. The reverse process is exfiltration – upward migration of water to the surface, or its vicinity, where the water enters the atmosphere by evaporation or transpiration through plants. Both flows are water flows through a partially saturated medium and in both the flow of two immiscible fluids is involved. When infiltration occurs, water flows into the soil and the air it displaces flows out. Here we are interested in the flow of the water, but the other should not be forgotten.

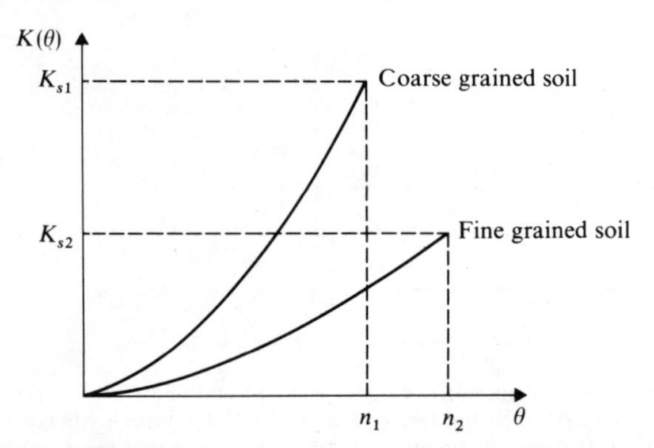

Fig. 1.16. $K(\theta)$ as a function of moisture content θ. When $\theta = n$, the porosity of the soil, $K = K_S$, the permeability of the saturated soil.

Percolation through a partially saturated medium is described by Darcy's law (section 1.9) and we have shown that gradients of capillary pressure act to drive the water through the soil. Both capillary pressure p_c (or the pressure head $\Psi = p_c/\gamma$) and the permeability or transmission coefficient K vary with moisture content in a given soil. As far as the flow of water is concerned, discontinuous pockets of air are part of the soil matrix. When the amount of air in the soil decreases, more passages are opened for the flow of water and K increases until the soil is saturated. This increase is shown diagrammatically in Fig. 1.16. In Fig. 1.17 some experimental results are shown, together with curves computed from measured properties of the soils. The abscissa for each of these curves is the saturation s, defined as the ratio of the volume of the water in the voids to the total volume of voids. The saturation $s = \theta/n$, where n is the porosity. The ordinates show two relative intrinsic permeabilities, k_{rw} for the wetting phase and k_{rnw} for the non-wetting phase. For infiltration, the wetting phase is water and the non-wetting phase is air. The intrinsic permeability k is defined by

$$K = k\gamma/\mu$$

as shown in equation 1.5, section 1.5 and the relative intrinsic permeability is the ratio of k at saturation s to k at $s = 1$.

The capillary pressure, like the permeability, depends on moisture content and the shape and size of the pores in the soil. However, it is not a single valued function and experimental results show hystersis, different curves for wetting and drying, Fig. 1.18. This hysteresis is the result of instability of the air-water

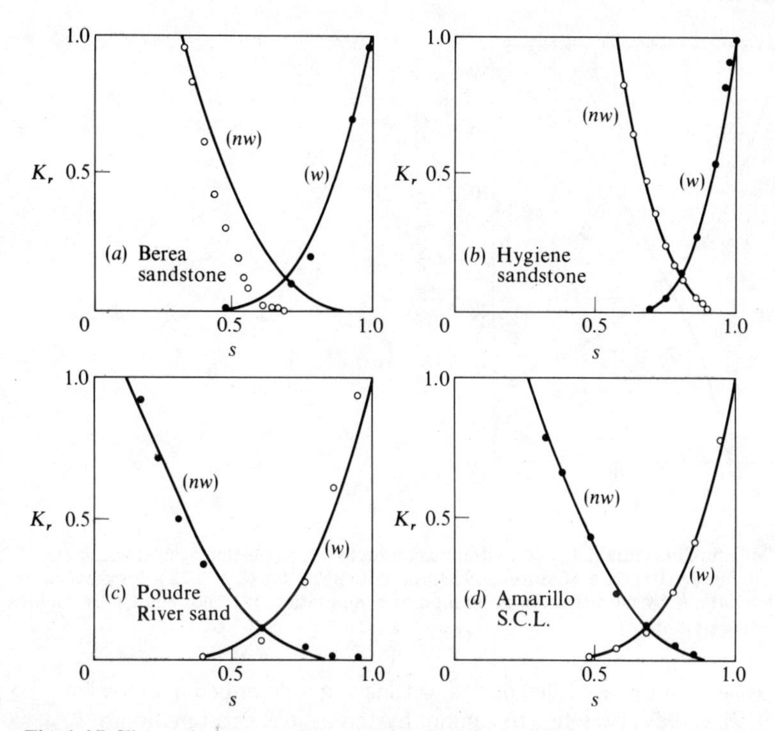

Fig. 1.17. Theoretical curves and experimental data for relative permeabilities of the non-wetting (*nw*) and wetting (*w*) phases as a function of saturation for two consolidated and two unconsolidated porous media, from Brooks and Corey, 1966.

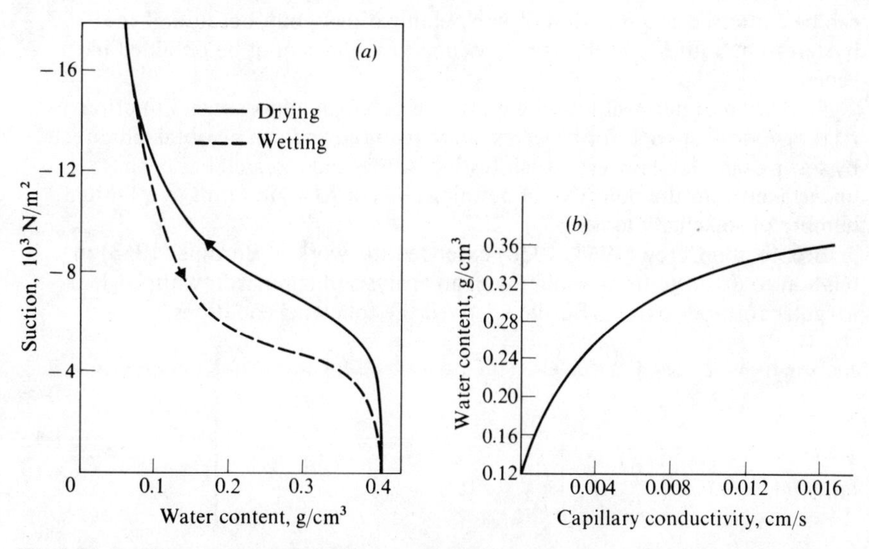

Fig. 1.18. Capillary suction and capillary conductivity as functions of moisture content. (*a*) Hysteresis effects on moisture-suction characteristics of a 50–500 μm sand sample, after Jackson, Reginato and van Bavel, 1965. (*b*) After Rogers and Klute, 1971.

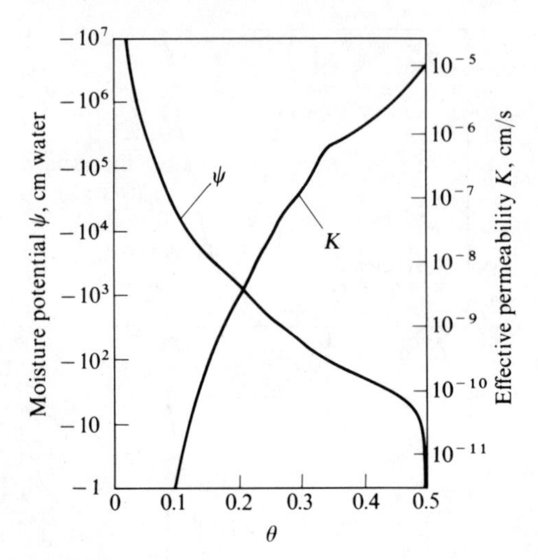

Fig. 1.19. 'Moisture potential', Ψ, and hydraulic conductivity, K, plotted against volumetric moisture content, θ, after data of Moore, 1939, for Yolo light clay ($\theta > 0.22$). Ψ extended on basis of other data, K by modification of method of computation of Childs and Collis–George, 1950, according to Philip, 1957.

interface as larger pores are filled during wetting and is described in detail by Childs (1969). Childs also refers to a minor hysteresis in K as a function of θ, attributed to the fact that different pores will be filled on different occasions when the moisture content is the same.

Typical curves for $\Psi(\theta)$ and $K(\theta)$ are shown in Fig. 1.19. The permeability K can be expressed as a function of Ψ by eliminating θ, but, because of the hysteresis in both K and Ψ, a single valued function cannot be obtained in general.

Prediction of permeability is a matter of practical importance. Functions relating K to θ or to Ψ for a given soil are required and can be obtained directly by testing samples. However, such testing is slow and research has been undertaken with the objective of defining $K(\theta)$ or $K(\Psi)$ in terms of a limited number of soil characteristics.

Brooks and Corey (1964, 1966) extended the work of Burdine (1953) in relation to drainage from a soil. From an analysis of laminar flow through an irregular tortuous passage Burdine derived the following equations:

$$\left.\begin{aligned} k_{rw} &= s_e^2 \int_0^{s_e} \frac{ds}{p_c^2} \Big/ \int_0^1 \frac{ds}{p_c^2} \\ k_{rnw} &= (1 - s_e^2) \int_{s_e}^1 \frac{ds}{p_c^2} \Big/ \int_0^1 \frac{ds}{p_c^2} \end{aligned}\right\} \qquad 1.15$$

Here s_e is the effective saturation defined by

$$s_e = (s - s_0)/(1 - s_0) \qquad 1.16$$

with s_0 the saturation corresponding to the residual moisture content θ_0. The relative intrinsic permeabilities are shown to depend on the variation of the capillary pressure p_c with the moisture content. Evaluation of k_{rw} and k_{rnw} is possible if p_c or Ψ can be defined as a function of s and Brooks and Corey (1964) showed by experiment that the following relationship holds for the soils they tested:

$$s_e = \begin{cases} (p_b/p_c)^{\lambda}, & -p_c \geqslant p_b \\ 1, & -p_c < p_b \end{cases} \tag{1.17}$$

The pressure p_b is the bubbling pressure and is the pressure at which the first continuous air-filled passage is formed as the suction on a water filled sample is increased. It is indicated by the appearance of bubbles in the testing apparatus. The bubbling pressure and the exponent λ are parameters which, like s_0 in equation 1.16 depend on the type of soil.

From equation 1.16 we obtain

$$s = (1 - s_0)\left(\frac{p_b}{p_c}\right)^{\lambda} - s_0 \tag{1.18}$$

This gives s as a function of p_c with parameters s_0, p_b and λ and substitution in equation 1.15 gives the relative intrinsic permeabilities in terms of the same parameters, as follows:

$$k_{rw} = s_e^{(2+3\lambda)\lambda} = \left(\frac{p_b}{p_c}\right)^{2+3\lambda} \tag{1.19}$$

A similar equation was obtained for the non-wetting phase and these are the theoretical curves in Fig. 1.17. It must be remembered that the parameters will have different numerical values for wetting and drying because of hysteresis and that Brooks and Corey dealt with drying.

Skaggs *et al.* (1970) took a function for $K(\Psi)$ proposed by Gardner (1958):

$$K(\Psi) = \left[\left(\frac{\Psi}{\Psi_1}\right)^{a} + b\right]^{-1} \tag{1.20}$$

They evaluated the parameters Ψ_1, a and b by comparing the results of infiltration tests on samples with computed solutions for the same flow and minimizing the difference between them. As we will show, Ψ must be known as a function of θ and Skaggs *et al.* used equation 1.17. The work of finding the desired values of Ψ_1, a and b was reduced by evaluating a and b first and using the minimum error criterion for Ψ_1 only. To find b, they used the fact that substituting $\Psi = 0$ in equation 1.20 yields.

$$K(0) = \frac{1}{b}$$

Thus, the value of K at saturation gives b. To find a, they used the fact that, for large values of Ψ

$$K(\Psi) \to \left(\frac{\Psi_1}{\Psi}\right)^{a}$$

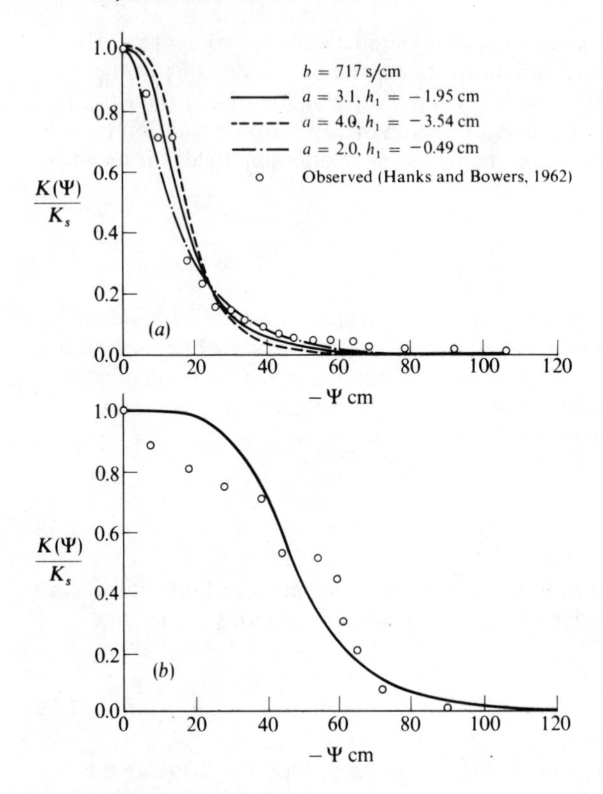

Fig. 1.20. Relative conductivity versus pressure head, (*a*) for Sarpy loam, (*b*) for Castor loam, according to Skaggs *et al.*, 1970.

This equation has the same form as equation 1.19 derived by Brooks and Corey. When observations are plotted on logarithmic scales, a is found from the slope of the straight line that fits them for large values of Ψ.

Fig. 1.20 shows some of the results obtained by Skaggs *et al.* In both cases, the full line is computed using optimal values of Ψ_1, a and b. In Fig. 1.20(a) two other curves are shown for off-optimum values of a and Ψ_1 for comparison.

The equations which describe infiltration are Darcy's law for the velocity vector and the continuity equation

$$\mathbf{q} = -K\nabla h$$

$$-\nabla \cdot \mathbf{q} = \frac{\partial \theta}{\partial t}$$

$$h = \frac{p}{\gamma} + z = \Psi + z$$

assuming constant density ρ and using $p = p_c$ for a partially saturated medium.

Hence

$$\frac{\partial\theta}{\partial t} = \nabla \cdot \{K\nabla(\Psi + z)\}$$

$$= \nabla \cdot \{K\nabla(\Psi) + K\mathbf{k}\}$$

$$= \nabla \cdot \{K\nabla(\Psi)\} + \frac{\partial K}{\partial z}$$

where

$$\nabla = \mathbf{i}\frac{\partial}{\partial x} + \mathbf{j}\frac{\partial}{\partial y} + \mathbf{k}\frac{\partial}{\partial z}$$

and

$$\nabla \cdot \mathbf{q} = \mathrm{div}\,\mathbf{q} = \frac{\partial u}{\partial x} + \frac{\partial v}{\partial y} + \frac{\partial w}{\partial z}$$

We can now use $K = K(\theta)$ and $\Psi = \Psi(\theta)$ to derive an equation in θ or $K = K(\Psi)$ and $\theta = \theta(\Psi)$, the inverse of $\Psi = \Psi(\theta)$, to derive one in Ψ. Because of hysteresis, these are not single-valued functions, some of them significantly so, and care is necessary to ensure that the correct functions are used.

In terms of θ we have

$$\frac{\partial\theta}{\partial t} = \nabla \cdot \{K(\theta)\nabla[\Psi(\theta)]\} + \frac{\partial}{\partial z}[K(\theta)]$$

$$= \nabla \cdot \{K(\theta)\frac{\mathrm{d}\Psi(\theta)}{\mathrm{d}\theta}\nabla(\theta)\} + \frac{\partial}{\partial z}[K(\theta)] \qquad 1.21$$

$$= \nabla \cdot \{D(\theta)\nabla(\theta)\} + \frac{\partial}{\partial z}[K(\theta)]$$

where

$$D(\theta) = K(\theta)\frac{\mathrm{d}\Psi(\theta)}{\mathrm{d}\theta}$$

This grouping was first used by Buckingham (1907) and is now known as the soil water diffusivity with dimensions $[L^2/T]$.

In terms of Ψ we have

$$\frac{\partial\theta}{\partial t} = \frac{\mathrm{d}\theta(\Psi)}{\mathrm{d}\Psi}\frac{\partial\Psi}{\partial t}$$

and

$$\frac{\mathrm{d}\theta}{\mathrm{d}\Psi}\frac{\partial\Psi}{\partial t} = \nabla \cdot \{K(\Psi)\nabla\Psi\} + \frac{\partial}{\partial z}[K(\Psi)] \qquad 1.22$$

Either of equations 1.21 or 1.22 may be solved numerically and closed solutions can be obtained if enough simplifying assumptions are made. For most soils the parameters $K(\Psi)$, $D(\theta)$ and $\mathrm{d}\theta/\mathrm{d}\Psi$ vary markedly with water content or the matrix suction head. The usual simplifying assumption is uniqueness of these relationships. Equation 1.21 in θ is preferred for numerical solution of unsaturated flow because changes in θ and D are two or three orders of magnitude smaller than corresponding changes in Ψ and $d\theta/d\Psi$. Round-off errors are smaller as a result. However, as θ approaches saturation ($s = 1, \theta = n$), the driving potential becomes independent of moisture content and $D(\theta)$ tends to infinity. The numerical calculation diverges. Hence, solutions from or through to saturation have to use equation 1.22. The latter for saturated conditions reduces to the Laplace equation with K constant and $d\theta/d\Psi$ zero.

If the infiltration is assumed to be a one-dimensional flow in the vertical direction, parallel to $0z$, we have

$$\frac{\partial\theta}{\partial t} = \frac{\partial}{\partial z}\left[D(\theta)\frac{\partial\theta}{\partial z} + K(\theta)\right]$$

This equation was derived by Richards (1931).

Fig. 1.21 shows diagrammatically the various zones of solution.

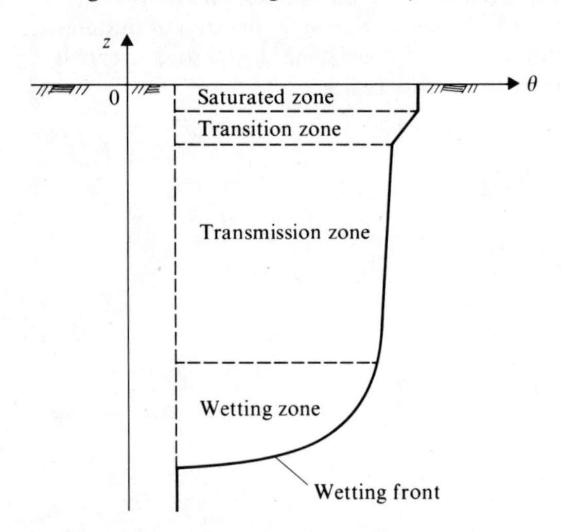

Fig. 1.21. Diagrammatic illustration of the zones of infiltration, after Bodman and Colman, 1943.

If K is approximately uniform or if gravitational effects are negligible, the second term on the right hand side can be dropped and we have equation 1.14:

$$\frac{\partial\theta}{\partial t} = \frac{\partial}{\partial z}\left[D(\theta)\frac{\partial\theta}{\partial z}\right] \qquad 1.14$$

These requirements are met approximately in the early stages of infiltration or late stages of exfiltration when the moisture content is low. The results obtained agree qualitatively with observation, except that hysteresis effects are underestimated. Equation 1.14 becomes the simple diffusion equation if a constant average diffusivity can be used in place of $D(\theta)$. Then

$$\frac{\partial\theta}{\partial t} = D\frac{\partial^2\theta}{\partial z^2} \qquad 1.23$$

An extensive treatment of this equation is available in Carslaw and Jaeger (1959) and Crank (1956).

The solution of equation 1.23 is determined by the boundary conditions. For example, if the rate of rainfall exceeds the infiltration capacity, the soil will be saturated continuously at the surface and for a semi-infinite slab of soil, the boundary conditions are

$$\theta = \begin{cases} \theta_1 & z \leq 0, t = 0 \\ \theta_2 & z = 0, t > 0 \end{cases}$$

Here θ_1 is the initial moisture content of the entire slab and $\theta_2 (= n$, the porosity) is the constant moisture content at the surface during infiltration. Equation 1.23 and these boundary conditions are satisfied by

$$\frac{\theta_2 - \theta}{\theta_2 - \theta_1} = \text{erf} \frac{|z|}{2\sqrt{Dt}}$$

At a depth $|z| = 4\sqrt{Dt}$, $\theta - \theta_1$ is negligible and this length is a measure of the influence of events at the surface. The total volume of infiltration through unit area after time t is

$$V_f = 2(\theta_2 - \theta_1)\sqrt{Dt/\pi}$$

and the rate of infiltration is given by the rate of diffusion at $z = 0$, namely

$$f_c = -D \left.\frac{\partial \theta}{\partial z}\right|_{z=0}$$

or

$$f_c = \left[-D \frac{\partial \theta}{\partial z} - K\right]_{z=0}$$

if gradients of K or gravitational effects or both cannot be neglected. Hence,

$$f_c = -(\theta_2 - \theta_1)\sqrt{D/\pi t} \qquad 1.24$$

or

$$f_c = -(\theta_2 - \theta_1)\sqrt{D/\pi t} - K_0 \qquad 1.25$$

where $K = K_0$ at $t = z = 0$.

Equations 1.24 and 1.25 show that the rate of infiltration decreases with time, and this accords with experience. However, at $t = 0$ they give $f_c = \infty$. This situation can be improved by solving equation 1.23 by separation of variables (the product method) and finding the velocity as a function of z and t. The result is

$$w = f = \sum_{m=0}^{m=\infty} (A \cos mz + B \sin mz) e^{-D^2 mt}$$

If

$$f = f_c, z = 0, t = \infty$$

$$f = f_i, z = 0, t = 0$$

then

$$f = f_c + (f_i - f_c)e^{-D^2mt}$$

which is Horton's equation.

Although it is implied in equation 1.21 that the infiltration type of flow is not limited to vertical movement of moisture, the point is emphazised here; some interesting flows in partially saturated porous media are approximately two-dimensional in vertical planes. One such flow is infiltration when the surface of the ground is sloping, as sketched on Fig. 1.22. At any location, there is a wetting front, with the various zones, which will penetrate the porous ground, as shown on Fig. 1.21. The saturated zone is sloping and a gradient of piezometric head drives the water downhill. This gradient exists because, for two points like A and B (Fig. 1.22), $p = 0$ and $\Delta h = \Delta z$. In the zones of unsaturated flow, the saturation θ will also usually have non-zero components $\partial\theta/\partial x$ and $\partial\theta/\partial z$ and they drive the unsaturated flow in a sloping direction. The flow is described by equation 1.21 with

$$\nabla = \mathrm{i}\frac{\partial}{\partial x} + \mathrm{k}\frac{\partial}{\partial z}$$

The expanded form of equation 1.21 then contains a lateral component, as well as the relationship for vertical infiltration.

Another similar kind of flow occurs when the surface of a lake or river is raised or lowered. Fully saturated flow of this kind is described in Chapter 8. The fringe of unsaturated flow above the wave of saturated flow is described by the two-dimensional version of equation 1.21, described above.

Methods for analysing such two-dimensional flows have been reviewed by Philip (1969) and Swartzendruber (1969).

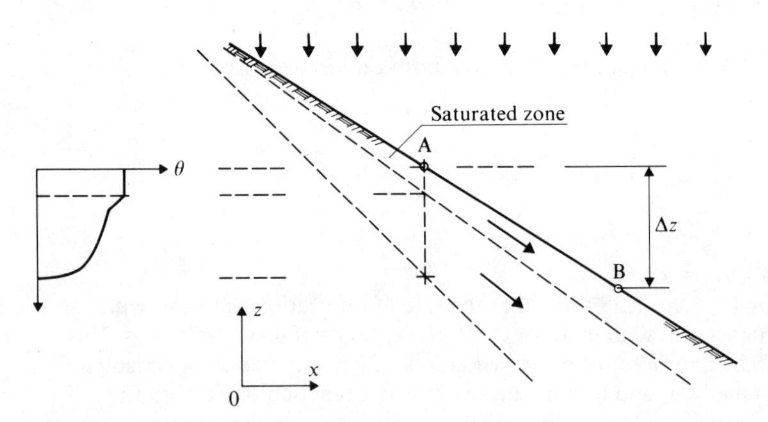

Fig. 1.22. Infiltration with sloping ground surface.

1.11 Diffusion and dispersion in groundwater

Pollution of groundwater is a problem of increasing concern to users of this resource. Transport and the ingress of solutes and other pollutants in porous media are discussed briefly in the following notes. Transport is affected by convection, by diffusion due to random motions of molecular and turbulent nature and the irregularities of the pores in the medium. Two approaches are in use. In one, a deterministic mathematical model is used, in the other a statistical model. For an extended discussion, Bear (1972) should be consulted.

Dispersion in a porous medium is made more complex than dispersion in other water flows because of the stochastic nature of the passages through which the water flows. On a microscopic scale the theories of hydrodynamic dispersion are valid. However, on a macroscopic scale, a control volume is an element of a porous medium and we have mechanical dispersion by the matrix of the medium in addition to convective, molecular and turbulent dispersion. The problem may be further complicated by adsorption of the contaminating substance at the solid surfaces of the medium and chemical and biochemical reactions may take place during transport.

The crux of the matter is evaluation of an effective diffusivity. When this is known problems can be solved by the same means as are used for conventional hydrodynamic diffusion. These have been described by Raudkivi and Callander (1975).

J. de Jong (1958) treated the porous medium as a network of straight channels. The channels were assumed to be of equal length and random orientation, the distribution with respect to direction being uniform. The average flow was at bulk velocity $\bar{V}$ in a specified direction. de Jong investigated the spreading of a tracer injected at a point in the flow. He obtained a three-dimensional normal distribution for the probability of arrival at a given point at time t. The corresponding surfaces of equal concentration are ellipsoids of revolution with their major axes parallel to the mean flow. The centre of this set of curves travels at speed $\bar{V}/3$ and the longitudinal and transverse standard deviations are given by

$$\sigma_L^2 = (l/3)(\lambda + 0.173)L$$

$$\sigma_T^2 = (3/8)lL$$

The corresponding diffusivities are

$$D_L = \sigma_L^2/2t$$

$$D_T = \sigma_T^2/2t$$

Here, the length $L = \bar{V}t$ and λ is given by $3L/l = (\lambda - 2.077)\exp(2\lambda)$. The reference length l is the length of each of the elementary channels in the network and is of the order of size of the grains in the medium.

Expressions for longitudinal and tranverse diffusivities were obtained by Saffman (1959, 1960) using a similar model. His results are as follows:

$$D_L = \frac{1}{2t}\overline{(x - \bar{V}t)^2} = \frac{1}{2}\,\bar{V}\,lS^2$$

$$S^2 = \ln(3\bar{V}\tau_0/l) - \frac{1}{12};\ \tau_0 = l^2/2D_m$$

$$\bar{V}\tau_0/l \leqslant [N\ln(3\,\bar{V}\tau_0/l)]^{1/2}$$

$$D_T = \frac{1}{2t}\,\overline{y^2} = \frac{3}{16}\,\bar{V}l$$

where $\bar{V}$ is the average bulk velocity and is parallel to Ox, l is the channel length; D_m is the molecular diffusivity and N is the number of channels traversed. It is

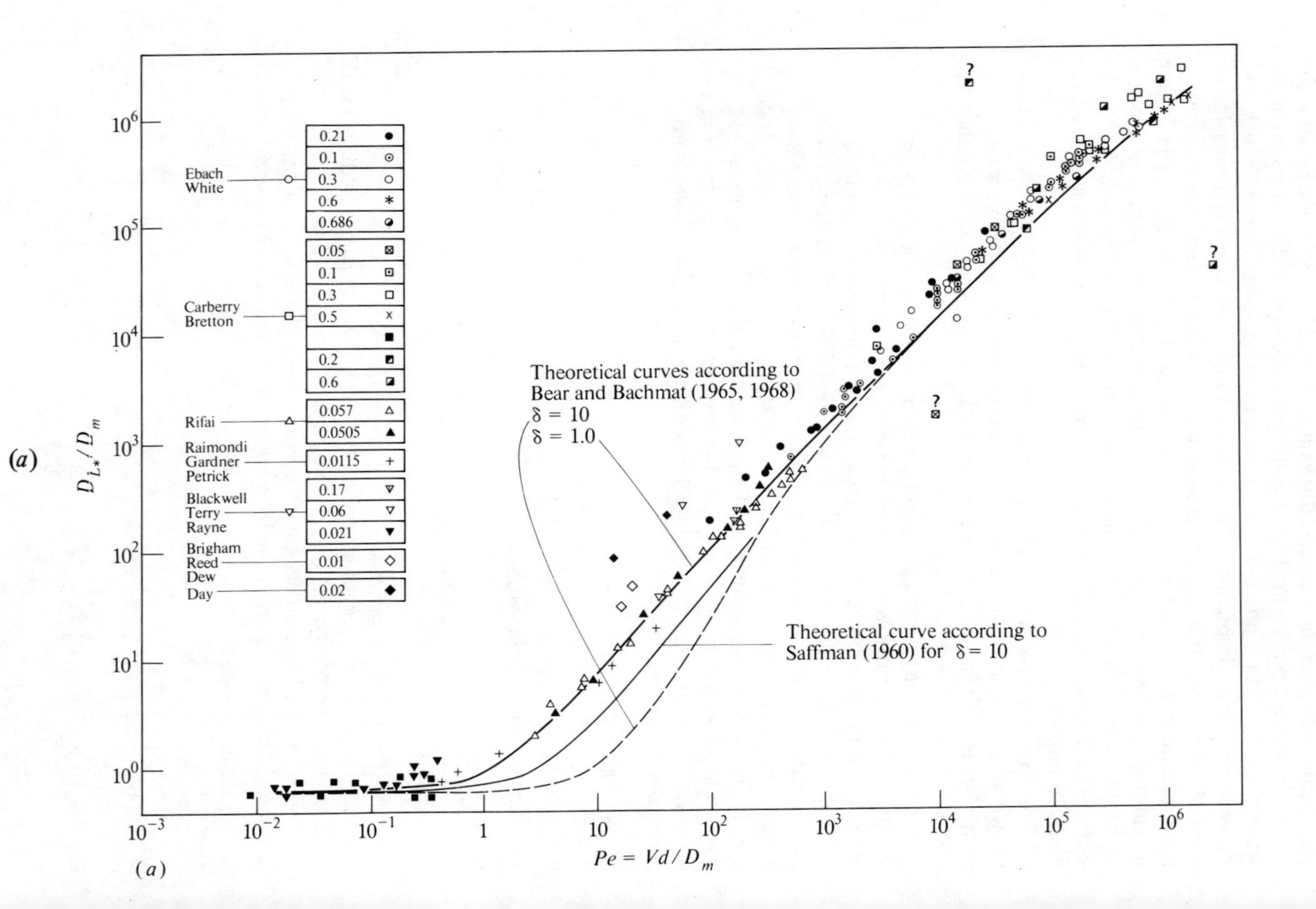

Ebach
White
0.21
0.1
0.3
0.6
0.686
Carberry
Bretton
0.05
0.1
0.3
0.5
0.2
0.6
Rifai
0.057
0.0505
Raimondi
Gardner
Petrick
0.0115
Blackwell
Terry
Rayne
0.17
0.06
0.021
Brigham
Reed
Dew
Day
0.01
0.02
Theoretical curves according to
Bear and Bachmat (1965, 1968)
δ = 10
δ = 1.0
Theoretical curve according to
Saffman (1960) for δ = 10
D_{L*}/D_m
$Pe = Vd/D_m$
(a)

(a)

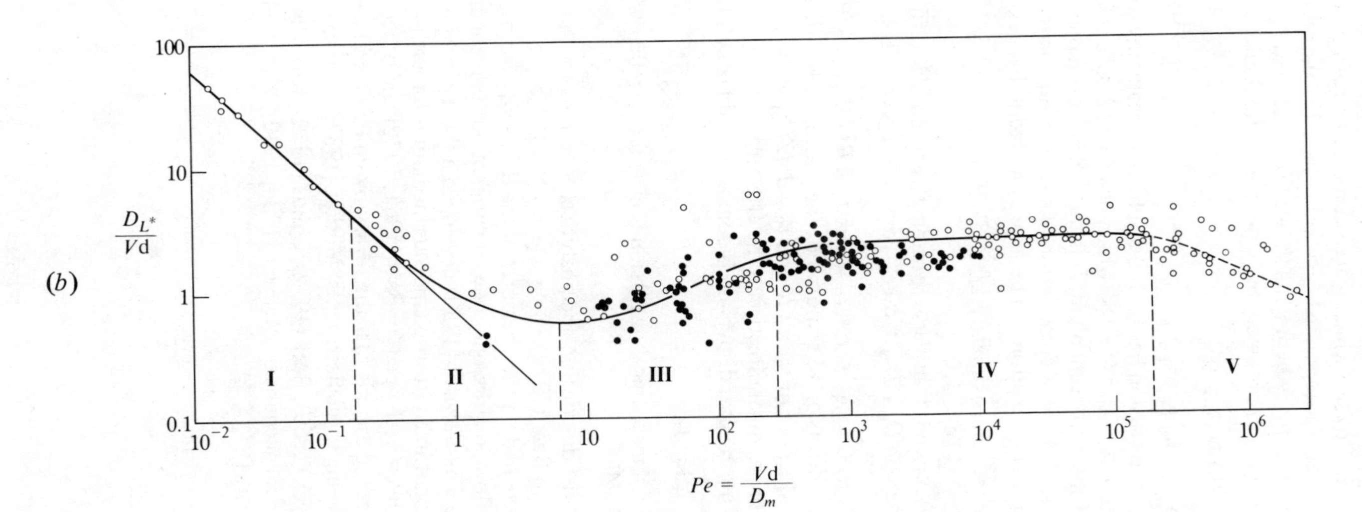

Fig. 1.23. (*a*) Relationship between molecular diffusion and convective dispersion, according to Pfannkuch, 1963, and Bear, 1972. (*b*) The regimes of dispersion according to Pfannkuch, 1963: I pure molecular diffusion, II superposition regime, III predominant mechanical dispersion, IV pure mechanical dispersion, V dispersion out of the Darcy domain. The exponent in $D_{L^*}/D_m = \alpha Pe^m$ is $m = 0$ for $P_e < 5$, $m = 1$ to 1.2 for $5 < Pe < 300$, $m = 1$ for $300 < Pe < 2 \times 10^5$ and $m < 1$ for $Pe > 2 \times 10^5$. D_{L^*} is the longitudinal sum of mechanical (convective) dispersion and molecular dispersion, δ characterizes the shape of the channels; it is the ratio of mean length between two adjacent junctions to a characteristic length of the pore cross-section, d is a measure of pore size assumed at times to be given by $\sqrt{k}$, and V is the average pore velocity.

necessary for N to be large enough to make the total distance travelled by marked particles statistically independent. The average value of N is $1.5\ \bar{V}t/l$.

Saffman analysed flows in which macroscopic and molecular mixing are of the same order of size. Extrapolated to a limit where $\bar{V} = 0$, this result showed $D_L = 1/3\ D_m$, whereas his experiments showed $D_L = 2/3\ D_m$.

Blackwell *et al.* (1959) obtained, by experiment, the result

$$D_L/D_m = 8.8\ \mathrm{Pe}^{1.7} \text{ for } \mathrm{Pe} > 0.5$$

where $\mathrm{Pe} = l\bar{V}/D_m$ is the Peclet number.

One-dimensional diffusion has been the subject of many investigations aimed at defining a diffusivity $D_* = D + D_m$ which depends on bulk velocity and the characteristics of the porous medium; D is the mechanical (convective) dispersion coefficient. Fig. 1.23 shows a compilation of such measurements as function of the Peclet number. The function can be approximated in five regions as follows:

(1) D_{L*}/D_m = const, approximately 2/3 as $\mathrm{Pe} \to 0$
(2) $D_{L*}/D_m \sim 1;\ 0 \leq \mathrm{Pe} < 5$
(3) A range where there is transverse spreading by molecular diffusion as well as longitudinal dispersion $D_{L*}/D_m = \alpha\ \mathrm{Pe}^m;\ \alpha \sim 0.5,\ 1 < m < 1.2$.

This expression is valid for $t_1 < t \ll t_0$ where $t_1 = R^2/8D_m$ is Saffman's time scale for molecular diffusion to smooth out the concentration distribution over the cross-section of a pore of radius R and $t_0 \propto d^2/2D_m$ is the time for molecular diffusion in the longitudinal direction to be effective.

(4) In this range, mechanical dispersion is dominant and Darcy's law is valid

$$D_{L*}/D_m = \beta\ \mathrm{Pe}, \quad \beta \sim 1.8$$

(5) A range in which mechanical dispersion is effective together with turbulent dispersion; $m < 1.74$.

For more detailed description and analytical functions for the relationship, reference is made to Bear (1972).

We have not touched on the character of the diffusivity for these flows. The one dimensional flows require only one component but two are derived in the model of de Jong. Nikolaevskii (1959) developed a theory in which the dispersion coefficient in an isotropic porous medium is a fourth-order tensor. These ideas have been developed by Bear (1961, 1972) and others.

For further reading there is first the review by Philip (1969) and reference is made to the following specialized texts: Muskat (1937), Polubarinova-Kochina (1962), Scheidegger (1960), Bear (1972), Childs (1969), Kirkham and Powers (1972), the historical account by Meinzer (1934, 1942), Walton (1970), and van der Leeden (1974).

2
Equation of continuity

The equation of continuity for flow through a porous medium is a partial differential equation which depends on conservation of mass. We work in terms of the discharge per unit area of the porous medium (the bulk velocity) and not the fluid velocity. Also, we will, in the first place, treat a general case in which the compressibility of the porous medium and the fluid are both allowed for. Introducing Darcy's law (essentially a force-momentum equation) enables us to write equations in which the bulk velocity is replaced by the piezometric head. Then we show that, with certain restrictions, the continuity equation can be reduced to the Laplace equation.

2.1 The equation of continuity

We consider the flow into and out of an elementary volume which in this case is an element of the porous medium, not an element of space. The net rate of flow into the element is

$$-\left[\frac{\partial}{\partial x}(\rho u)+\frac{\partial}{\partial y}(\rho v)+\frac{\partial}{\partial z}(\rho w)\right]\Delta x\Delta y\Delta z$$

where u, v and w are the components of the bulk velocity and ρ is the density of the fluid, and must equal the rate at which matter is being stored in the element

$$\frac{\partial}{\partial t}(n\rho\Delta x\Delta y\Delta z) \qquad 2.1$$

where n is the porosity.

If the porous medium is incompressible, n is constant and the element under consideration will not change in size. Then

$$\frac{\partial}{\partial x}(\rho u)+\frac{\partial}{\partial y}(\rho v)+\frac{\partial}{\partial z}(\rho w)=-n\frac{\partial\rho}{\partial t} \qquad 2.2$$

If the porous medium is compressible, the porosity of the medium may vary with position and time, and the dimensions of an element at any chosen position may vary with time. With regard to the latter, it is assumed that changes in the lateral dimensions Δx and Δy are constrained by the surrounding soil and that only changes in the vertical dimension Δz need be taken into account. Equation 2.1 may now be expanded as follows:

$$\frac{\partial}{\partial t}(n\rho\Delta x\Delta y\Delta z = \left[n\rho \frac{\partial}{\partial t}(\Delta z) + \rho\Delta z \frac{\partial n}{\partial t} + \Delta z\, n \frac{\partial \rho}{\partial t} \right] \Delta x \Delta y \qquad 2.3$$

We now simplify by introducing the compressibilities of the medium and the fluid. It is customary in this field to use the compressibility instead of the modulus of elasticity; they are reciprocals. For the soil, with vertical component of intergranualar pressure σ_{zz}, we have

$$E_s = \frac{1}{\alpha} = -\frac{\mathrm{d}\sigma_{zz}}{\left[\frac{\mathrm{d}(\Delta z)}{\Delta z}\right]}$$

i.e.

$$\mathrm{d}(\Delta z) = -\alpha\Delta z\, \mathrm{d}(\sigma_{zz})$$

and

$$\frac{\partial}{\partial t}(\Delta z) = -\alpha\Delta z \frac{\partial(\sigma_{zz})}{\partial t} \qquad 2.4$$

Thus, the first term on the right side of equation 2.3 becomes

$$-\alpha n\rho\Delta z \frac{\partial}{\partial t}(\sigma_{zz}) \qquad 2.5$$

With regard to the second term, we consider the volume of solids in the element to be constant, that is, the change in the volume of the element is brought about by a change in the volume of the pores, i.e.

$$(1-n)\,\Delta x\Delta y\Delta z = \text{constant} \qquad 2.6$$

whence, with Δx and Δy unchanging

$$\frac{\partial n}{\partial t} = \frac{1-n}{\Delta z}\frac{\partial}{\partial t}(\Delta z)$$

$$= -(1-n)\alpha \frac{\partial}{\partial t}\sigma_{zz} \qquad 2.7$$

The second step follows from equation 2.4 and the second term on the right side of equation 2.3 becomes

$$-\rho\,(1-n)\,\alpha\Delta z \frac{\partial\sigma_{zz}}{\partial t} \qquad 2.8$$

For the third term, we consider conservation of the matter of the fluid to obtain

$$\rho\Delta V = \rho_0\, \Delta V_0 = \text{constant} \qquad 2.9$$

where V is volume of fluid. Hence,

$$\frac{\mathrm{d}(\Delta V)}{\Delta V} + \frac{\mathrm{d}\rho}{\rho} = 0$$

Thus

$$\mathrm{d}\rho = -\rho\,\frac{\mathrm{d}(\Delta V)}{\Delta V}$$

The fluid compressibility

$$\beta = \frac{1}{E_w} = \frac{\mathrm{d}V}{V\mathrm{d}p}$$

where p is the fluid pressure so that

$$\mathrm{d}\rho = \beta\rho\,\mathrm{d}p \tag{2.10}$$

and

$$\frac{\partial\rho}{\partial t} = \beta\rho\,\frac{\partial p}{\partial t} \tag{2.11}$$

Finally, we make use of the approximation that the sum of σ_{zz} and p equals the weight of the overburden, so that $\sigma_{zz} + p$ = constant, with respect to time and

$$\frac{\partial\sigma_{zz}}{\partial t} = -\,\frac{\partial p}{\partial t} \tag{2.12}$$

Substituting in equation 2.3 and simplifying yields

$$\frac{\partial}{\partial x}(\rho u) + \frac{\partial}{\partial y}(\rho v) + \frac{\partial}{\partial z}(\rho w) = -\rho\beta\left(n + \frac{\alpha}{\beta}\right)\frac{\partial p}{\partial t} \tag{2.13}$$

We now replace the fluid pressure p by the piezometric head h using

$$h = \frac{p}{\gamma} + z$$

whence

$$p = \gamma h - \gamma z$$

Differentiating and remembering that z is not a function of x, y or t, we obtain

$$\frac{\partial p}{\partial x} = \rho g\,\frac{\partial h}{\partial x} + gh\,\frac{\partial\rho}{\partial x} - gz\,\frac{\partial\rho}{\partial x} = \rho g\,\frac{\partial h}{\partial x} + \frac{p}{\rho}\,\frac{\partial\rho}{\partial x}$$

$$\frac{\partial p}{\partial y} = \rho g\,\frac{\partial h}{\partial y} + \frac{p}{\rho}\,\frac{\partial\rho}{\partial y}$$

$$\frac{\partial p}{\partial z} = \rho g\left(\frac{\partial h}{\partial z} - 1\right) + (gh - gz)\frac{\partial\rho}{\partial z} = \rho g\left(\frac{\partial h}{\partial z} - 1\right) + \frac{p}{\rho}\,\frac{\partial\rho}{\partial z} \tag{2.14}$$

$$\frac{\partial p}{\partial t} = \rho g\,\frac{\partial h}{\partial t} + \frac{p}{\rho}\,\frac{\partial\rho}{\partial t}$$

Equation 2.14, with equation 2.11 gives

$$\frac{\partial \rho}{\partial x} = \beta\rho \frac{\partial p}{\partial x} = \beta\rho \left(\rho g \frac{\partial h}{\partial x} + \frac{p}{\rho} \frac{\partial \rho}{\partial x} \right)$$

$$\frac{\partial \rho}{\partial y} = \beta\rho \frac{\partial p}{\partial y} = \beta\rho \left(\rho g \frac{\partial h}{\partial y} + \frac{p}{\rho} \frac{\partial \sigma}{\partial y} \right)$$

$$\frac{\partial \rho}{\partial z} = \beta\rho \frac{\partial p}{\partial z} = \beta\rho \left[\rho g \left(\frac{\partial h}{\partial z} - 1 \right) + \frac{p}{\rho} \frac{\partial \rho}{\partial y} \right]$$

$$\frac{\partial \rho}{\partial t} = \beta\rho \frac{\partial p}{\partial t} = \beta\rho \left(\rho g \frac{\partial h}{\partial t} + \frac{p}{\rho} \frac{\partial \rho}{\partial t} \right)$$

Solving these for $\frac{\partial \rho}{\partial x}, \frac{\partial \rho}{\partial y}, \frac{\partial \rho}{\partial z}, \frac{\partial \rho}{\partial t}$ we obtain

$$\frac{\partial \rho}{\partial x} = \frac{1}{1 - \beta p} \beta\rho^2 g \frac{\partial h}{\partial x}$$

$$\frac{\partial \rho}{\partial y} = \frac{1}{1 - \beta p} \beta\rho^2 g \frac{\partial h}{\partial y}$$

$$\frac{\partial \rho}{\partial z} = \frac{1}{1 - \beta p} \beta\rho^2 g \left(\frac{\partial h}{\partial z} - 1 \right) \tag{2.15}$$

$$\frac{\partial \rho}{\partial t} = \beta\rho \frac{\partial p}{\partial t}$$

Also

$$\frac{\partial p}{\partial t} = \frac{1}{1 - \beta p} \rho g \frac{\partial h}{\partial t}$$

using the last of equation 2.14.

Since $E_w = 2.07$ GPa and $\beta \simeq 4.829 \times 10^{-10} \text{m}^2\text{N}^{-1}$, $1/(1 - \beta p) = 10^9/(10^9 - 0.48p) \simeq 1$ for a wide range of pressures.

Returning now to equation 2.13, the left side may be written as

$$\rho \left(\frac{\partial u}{\partial x} + \frac{\partial v}{\partial y} + \frac{\partial w}{\partial z} \right) + u \frac{\partial \rho}{\partial x} + v \frac{\partial \rho}{\partial y} + w \frac{\partial \rho}{\partial z}$$

and by equation 1.11, 1.12, the first group of terms becomes

$$-\rho K \left(\frac{\partial^2 h}{\partial x^2} + \frac{\partial^2 h}{\partial y^2} + \frac{\partial^2 h}{\partial z^2} \right) = -\rho K \nabla^2 h = -\rho \nabla^2 \phi \tag{2.16}$$

Note that, although the permeability does in fact vary with the over-burden pressure, this dependence has been ignored (Childs and Collis-George, 1950).

The terms containing the partial derivatives of the density give, on substitution from equation 2.15 and equation 1.11,

$$-\beta\rho^2 gK\left[\left(\frac{\partial h}{\partial x}\right)^2+\left(\frac{\partial h}{\partial y}\right)^2+\left(\frac{\partial h}{\partial z}\right)^2-\frac{\partial h}{\partial z}\right] \qquad 2.17$$

The sum of the two expressions 2.16 and 2.17 equals, approximately, the right side of equation 2.13, i.e.

$$-\beta\rho^2 g\left(n+\frac{\alpha}{\beta}\right)\frac{\partial h}{\partial t}$$

Generally, the terms in equation 2.17 are small compared with those in equation 2.16. This may be seen from consideration of the physical significance of the two groups. In equation 2.16, we have the product of density and velocity gradients, in equation 2.17, the product of density gradients and velocities. Since the velocities, velocity gradients and density gradients are all small quantities and density is not – especially for groundwater flow – equation 2.17 is a second order small quantity.

Thus, with good approximation the continuity equation reduces to

$$\nabla^2 h=\frac{\beta\rho g\left(n+\dfrac{\alpha}{\beta}\right)}{K}\frac{\partial h}{\partial t} \qquad 2.18$$

The numerator of the factor on the right side is known as the specific storage [1/L]; $\rho g\alpha$ gives the water yield from storage as a result of compression of the porous medium and $\beta\rho g n$ is the yield resulting from a corresponding expansion of the water. For partially saturated soils the compressibility terms become unimportant and the right-hand side of equation 2.18 becomes $\partial\theta/\partial t$.

2.2 Two-dimensional flows

For steady flows, $\partial h/\partial t = 0$ and equation 2.18 reduces to the Laplace equation

$$\nabla^2 h = 0$$

However, with h a function of the three space coordinates x, y and z, to solve the equation is a considerable task. By contrast, certain two-dimensional flows can be dealt with quite simply. What we seek to do now is to show that the continuity equation for these flows reduces to the Laplace equation in two dimensions.

Firstly, if the y component of bulk velocity, v, is zero, h must be a function of x and z and the continuity equation becomes

$$\nabla^2 h=\frac{\partial^2 h}{\partial x^2}+\frac{\partial^2 h}{\partial z^2}=0$$

Or, using $\phi = Kh$ + constant, with K constant

$$\nabla^2\phi = 0$$

The flow of groundwater through vertical slices of the porous medium is analogous to a potential flow, and the flow net can be found by solving the Laplace equation as if the problem were one involving potential flow.

Secondly, if the z component of velocity, w, is zero, and u and v can be assumed unvarying with respect to z, h can be found as a function of x and y.

We derive the continuity equation for an incompressible medium and an incompressible fluid by considering the flow into and out of an element of the medium, as before. This time, however, the height of the element Δz is the full height of the aquifer. It is, of course, not now infinitesimal. The lower boundary of the element is an impervious surface and the upper boundary is either another impervious surface or the water table. We will find two important special cases in which the continuity equation again is reduced to the two-dimensional Laplace equation.

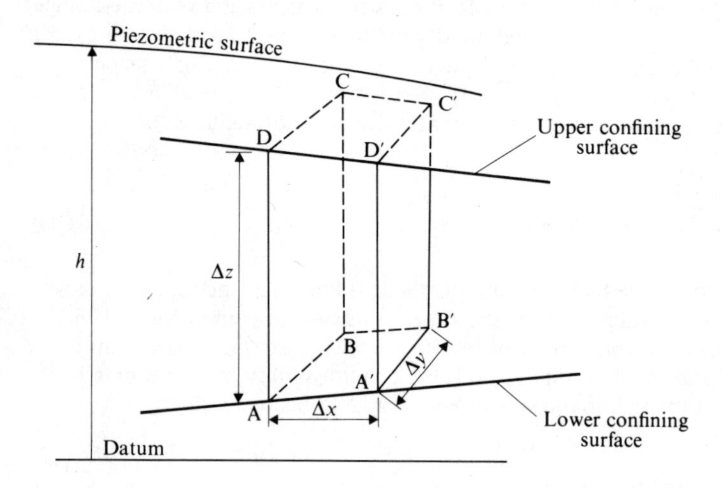

Fig. 2.1. Application of continuity equation to flow in a confined aquifer.

From Fig. 2.1:

Inflow to element through ABCD $= u\Delta y\Delta z$

Outflow through A′B′C′D′ $= u\Delta y\Delta z + (\partial/\partial x)(u\Delta y\Delta z)\Delta x$ and the net outflow from the element through these surfaces

$$\frac{\partial}{\partial x}(u\Delta y\Delta z)\Delta x = \frac{\partial}{\partial x}(u\Delta z)\Delta x\Delta y$$

since Δy is constant. Similarly the net outflow through the faces AA′D′D and BB′C′C $= (\partial/\partial y)(v\Delta z)\Delta x\Delta y$. The total net outflow must be zero, since there is no change in the volume of water stored in the element. Therefore, as Δx and Δy tend to zero

$$\frac{\partial}{\partial x}(u\Delta z) + \frac{\partial}{\partial y}(v\Delta z) = 0 \qquad 2.19$$

If we now make Δz = constant, i.e. the flow is between parallel impermeable surfaces, we have

$$\frac{\partial u}{\partial x} + \frac{\partial v}{\partial y} = 0$$

and substituting

$$u = -K \frac{\partial h}{\partial x}$$

$$v = -K \frac{\partial h}{\partial y}$$

we obtain

$$\frac{\partial^2 h}{\partial x^2} + \frac{\partial^2 h}{\partial y^2} = 0$$

A potential function $\phi = Kh$ + constant satisfies the equation, if K is a constant, that is $\nabla^2 \phi = 0$.

For the second special case the upper surface of the flow is the water table.

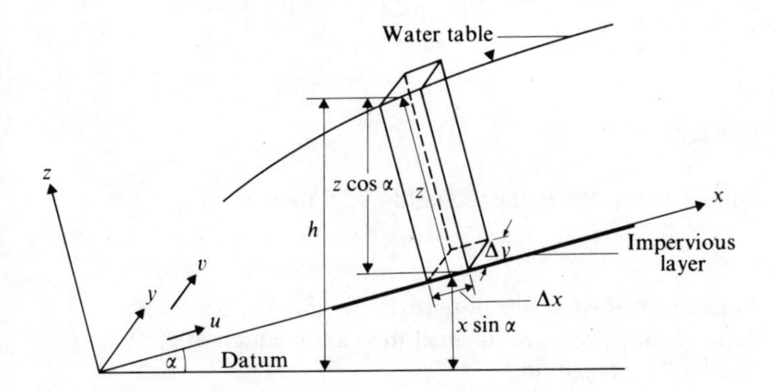

Fig. 2.2. Application of continuity equation to unconfined flow through a porous medium.

With the symbols of Fig. 2.2 the continuity equation for steady flow is

$$\frac{\partial}{\partial x}(zu)\mathrm{d}x\mathrm{d}y + \frac{\partial}{\partial y}(zv)\mathrm{d}y\mathrm{d}x = 0$$

or

$$\frac{\partial}{\partial x}(zu) + \frac{\partial}{\partial y}(zv) = 0$$

The piezometric head is

$$h = z \cos \alpha + x \sin \alpha$$

and by the Darcy law

$$u = -K \frac{\partial h}{\partial x}$$

$$v = -K \frac{\partial h}{\partial y}$$

where

$$\frac{\partial h}{\partial x} = \frac{\partial z}{\partial x} \cos \alpha + \sin \alpha$$

$$\frac{\partial h}{\partial y} = \frac{\partial z}{\partial y} \cos \alpha$$

Substituting these into the continuity equation yields

$$\frac{\partial}{\partial x}\left[-Kz\left(\frac{\partial z}{\partial x} \cos \alpha + \sin \alpha\right)\right] + \frac{\partial}{\partial y}\left[-Kz \frac{\partial z}{\partial y} \cos \alpha\right] = 0$$

$$\frac{\partial}{\partial x}\left[-Kz \frac{\partial z}{\partial x} \cos \alpha\right] + \frac{\partial}{\partial y}\left[-Kz \frac{\partial z}{\partial y} \cos \alpha\right] + \frac{\partial}{\partial x}\left[-Kz \sin \alpha\right] = 0$$

or

$$\nabla^2 (\tfrac{1}{2}Kz^2) + \frac{\partial}{\partial x}(Kz \tan \alpha) = 0 \qquad 2.20$$

If the lower confining layer is horizontal, equation 2.20 reduces to

$$\nabla^2 (\tfrac{1}{2}Kz^2) = 0$$

where $\frac{1}{2}Kz^2$ could be indentified with a potential function, i.e. $\phi = \frac{1}{2}Kz^2$ + constant. If the slope of the lower layer is small then again equation 2.20 approximates to the Laplace equation.

These special cases of groundwater flow are important because of the existence of solutions to analogous potential flow problems. The analogy enables us to use these solutions to find the distribution of piezometric head for groundwater flows. Again we emphasize that flow through a porous medium is not an irrotational flow; what we are dealing with is an analogy. Furthermore, the restriction imposed by the assumptions must be remembered; the flow is steady, the bulk velocity is assumed not to vary with z, Darcy's law is valid, the permeability is constant and compressibility of the porous medium and the fluid are negligible. If the flow is in the inertia range, $(\partial h/\partial x) \propto U^2$, etc. and even though all the other conditions are satisfied, the differential equation derived from continuity will be non-linear.

In the following chapters, attention will be confined to flows which can be assumed to be two-dimensional, that is, flows in vertical slices or between parallel confining layers or free surface flows. We will use the linear nature of the Laplace equation to superimpose elementary solutions in building up the solutions to more complex problems. The restrictions mentioned above which enable the continuity equation to be reduced to the Laplace equation will be applied. However, the preceding discussion of the wider field of study should not be overlooked and we hope that it will serve to show the following chapters correctly in relation to the wider field.

3 Steady confined flow

The foregoing general discussion can now be applied to the solution of some problems of flow through a porous medium confined between parallel impervious surfaces, not necessarily horizontal. We will consider two elementary problems first – a uniform flow through the medium, analogous to a uniform parallel potential flow, and radial flow to a well which is analogous to flow to a sink in potential flow. Then, several combinations of two or more of these elementary flows will be considered. The analogy with two-dimensional potential flow will be used in vertical slices or in the x-y plane, depending on which is more convenient for the problem in hand. When it seems desirable to emphasize the groundwater flow, rather than the analogy, we will derive the equation from Darcy's law.

In all cases, our objective is to find the distribution of the piezometric head as a function of the space coordinates. The distributions depend on the permeability of the porous medium and on the discharge, so that these quantities appear as parameters in the equations for h. It follows that only minor rearrangement is required to get K or Q from observations of piezometric head.

3.1 Uniform flow

For the analogous potential flow, the potential is given by

$$\phi = -U_0 x \qquad 3.1$$

The minus sign arises from the definition that a positive potential gradient causes a velocity in the negative direction. Because the analogy exists, the same equation gives the groundwater velocity potential defined by

$$\phi = Kh + C \qquad 3.2$$

Hence

$$h = -\frac{U_0}{K}x - C \qquad 3.3$$

The constant C depends on the choice of datum for h and can conveniently be made to disappear if $h = 0$ when $x = 0$. Then

$$h = -\frac{U_0}{K}x \qquad 3.4$$

The equipotentials for the groundwater flow can be plotted using equation 3.1. They are the parallel straight lines

$$x = -\frac{\phi}{U_0} \qquad 3.5$$

The y axis is chosen arbitrarily to represent $\phi = 0$.

To complete the flow net, the concept of a stream function can be used. The stream function for the groundwater flow, analogous to the potential flow, is

$$\psi = -U_0 y \qquad 3.6$$

the line $y = 0$ being chosen arbitrarily to represent the streamline $\psi = 0$. The streamlines are given by

$$y = -\frac{\psi}{U_0} \qquad 3.7$$

Thus, the flow net is the rectangular mesh described by equations 3.5 and 3.7. The flow net can, of course, be found directly without using the analogy.

With no component of velocity parallel to the y axis, the continuity equation may be simplified as follows:

$$\nabla^2 h = \frac{\partial^2 h}{\partial x^2} = \frac{\mathrm{d}^2 h}{\mathrm{d}x^2} = 0 \qquad 3.8$$

Integration yields

$$\frac{\mathrm{d}h}{\mathrm{d}x} = C_1$$

and

$$h = C_1 x + C_2$$

We also have from Darcy's law

$$-K\frac{\mathrm{d}h}{\mathrm{d}x} = u = U_0$$

so that

$$C_1 = -\frac{U_0}{K}$$

With an appropriate choice of datum, $C_2 = 0$ and, as before,

$$h = \frac{-U_0}{K}x \qquad 3.4$$

3.2 Radial flow to a well

Radial flow to a fully penetrating well is a two-dimensional flow which is symmetrical with respect to the axis of the well. The properties of the flow vary only with radial distance from the well, so that they are most readily described in polar coordinates. The solution can be arrived at either by using the analogous potential flow (flow to a sink) or by solving the continuity equation.

The problem we solve here is about flow to a well through an aquifer which is circular in plan. The well is at the centre of the circle and at the outer edge of the aquifer a constant head of water is maintained. Fig. 3.1 is a vertical section through the centre of the system and shows the definitions of the symbols used.

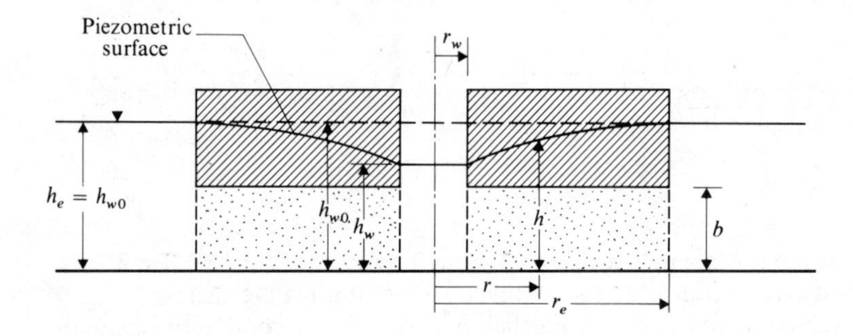

Fig. 3.1. Definition diagram for radial flow in a confined aquifer of constant thickness.

For potential flow to a sink,

$$\phi = m \ln r \tag{3.9}$$

Here, m is the strength of the sink and $r_w < r < r_e$. For the groundwater flow the same equation is valid with

$$\phi = Kh + C \tag{3.2}$$

and

$$Kh = m \ln r - C \tag{3.10}$$

We have now to evaluate m and C in terms of suitable parameters of the groundwater flow. The boundary conditions which are used for this purpose are

$h = h_w$ when $r = r_w$

$h = h_e$ when $r = r_e$

where the water level in the well h_w must always be above the upper confining layer. The boundary conditions give

$$Kh_w = m \ln r_w - C$$

and

$$Kh_e = m \ln r_e - C$$

$$m = \frac{K(h_e - h_w)}{\ln \dfrac{r_e}{r_w}} \tag{3.11}$$

and

$$C = K \frac{h_e - h_w}{\ln \dfrac{r_e}{r_w}} \ln r_w - Kh_w$$

Substituting in equation 3.10 yields

$$h - h_w = (h_e - h_w) \frac{\ln \dfrac{r}{r_w}}{\ln \dfrac{r_e}{r_w}}$$

or

$$\frac{h - h_w}{h_e - h_w} = \frac{\ln \dfrac{r}{r_w}}{\ln \dfrac{r_e}{r_w}} \tag{3.12}$$

The discharge to the well can be related to the strength of the sink in the potential flow. We have, for the stream function of the sink,

$$\psi = m\theta \tag{3.13}$$

The total flow to the sink is $2\pi m$ and this is analogous to the discharge to the well through unit thickness of the aquifer, $q = Q/b, Q$ being the total discharge. Hence

$$m = \frac{q}{2\pi} \tag{3.14}$$

Equations 3.11 and 3.14 give us

$$q = \frac{2\pi K(h_e - h_w)}{\ln \dfrac{r_e}{r_w}} \tag{3.15}$$

which enables the discharge to be found when the coefficient of permeability and the geometry of the system are known.

Equations 3.9 and 3.14 yield for the groundwater velocity potential

$$\phi = \frac{q}{2\pi} \ln r$$

Hence, $\phi = 0$ when $r = 1$. Since it is more convenient to have $\phi = 0$ when $r = r_w$, we shift the datum for ϕ and make

$$\phi = \frac{q}{2\pi} \ln \frac{r}{r_w} \qquad 3.16$$

This shift of datum does not affect any of the conclusions derived above, except that C will have a different value.

Similarly, equation 3.13 and 3.14 give for the stream function

$$\psi = \frac{q}{2\pi} \theta \qquad 3.17$$

Equations 3.16 and 3.17 define the flow net, a set of circles for the equipotentials and a set of radial straight lines for the streamlines.

Various rearrangements of these equations can be written, to derive useful formulas. For example, what is known as the 'drawdown', $h_e - h$ can be used to describe the water table instead of $h - h_w$ as in equation 3.12:

$$\frac{h_e - h}{h_e - h_w} = \frac{\ln \frac{r_e}{r}}{\ln \frac{r_e}{r_w}} \qquad 3.18$$

or, using equation 3.15

$$h_e - h = \frac{q}{2\pi K} \ln \frac{r_e}{r} \qquad 3.19$$

At the well face, the drawdown is

$$h_e - h_w = \frac{q}{2\pi K} \ln \frac{r_e}{r_w} \qquad 3.20$$

Another rearrangement leads to an expression for K which is useful in determination of the permeability of the aquifer. If the piezometric head is measured in two observation wells at radii r_1 and r_2 and found to be h_1 and h_2, we have

$$K = \frac{q}{2\pi(h_2 - h_1)} \ln \frac{r_2}{r_1} \qquad 3.21$$

The foregoing results are valid for the particular problem postulated and we now need to consider whether they can be used as approximations when there is no surrounding reservoir. In this case, the boundary condition $h = h_e$ when $r = r_e$ is not available and another condition must be found to replace it if the constants m and C (equation 3.10) are to evaluated. The best that can be done is to use field measurements to find a notational external radius which is assumed to be valid for all other flows as well as the test flow. If a pumping test is performed there are two possibilities:

(1) The discharge $Q = qb$ and the drawdown $(h_e - h_w)$ are measured. Equation 3.15 may then be used to calculate r_e.

(2) The piezometric head h is measured in an observation well at radius r (preferably large). Since h_e and h_w are also known, equation 3.12 enables r_e to be calculated.

In both cases r_e is defined by the intersection of the logarithmic drawdown curve with $h = h_e$ which is the undisturbed piezometric surface. If the test flow is large, with a substantial drawdown at the well, the value of r_e obtained should enable discharge and drawdown profiles to be predicted within reasonable limits.

When it is not possible to make field measurements, the second boundary condition must be guessed and it is better to guess the external radius r_e than the discharge. Since the equations contain the logarithm of r_e the errors in discharge and drawdown arising from inaccurate guessing are much smaller than the error in the estimate of r_e. Judicious interpretation of local knowledge is required.

The point we emphasize here is that an alternative to the real boundary condition $h = h_e$ when $r = r_e$ has to be found, and this will be done most satisfactorily by means of a field test.

It may also be noted that since the discharge varies with the logarithms of the radii of the well and the exterior boundary, very large variations in these radii are required to cause an appreciable change in discharge. For example, if r_e/r_w is 1600, a fortyfold increase in r_w is required to double the discharge.

All the above results can, of course, be derived directly from Darcy's law without recourse to the potential flow analogy. Using plane polar coordinates, we have

$$v_\theta = -K \frac{1}{r} \frac{\partial h}{\partial \theta} = 0$$

and

$$v_r = -K \frac{\partial h}{\partial r} = -K \frac{dh}{dr} \qquad 3.22$$

But

$$q = 2\pi r \, v_r$$

and

$$-K \frac{dh}{dr} = \frac{q}{2\pi} \frac{1}{r} \qquad 3.23$$

On integration, this yields,

$$-Kh = \frac{q}{2\pi} \ln r + C \qquad 3.24$$

and, inserting the boundary conditions

$$h = h_w \qquad \text{when} \qquad r = r_w$$

and

$$h = h_e \qquad \text{when} \qquad r = r_e$$

we obtain all the foregoing results.

Flow to a horizontal drain is represented by a vertical slice of the aquifer with the horizontal drain embedded to half its diameter in the confining layer and is analogous to flow into a sink. The potential flow analogy is again available with

$$\phi = Kh + C \qquad 3.2$$

Provided the aquifer is approximately semi-finite the flow net near the drain will be the same in the x, z plane as that for flow to a well in the x, y plane. One pair of radial streamlines coincides with the boundary of the aquifer near the drain.

3.3 Well in a uniform flow

Both the velocity potential ϕ and the stream function ψ satisfy the Laplace equation (Raudkivi and Callander (1975)). Since the Laplace equation is a linear equation, linear combinations of functions which are solutions are also solutions and two or more solutions can be superimposed on each other by simply adding the distributions of ϕ or ψ as the case may be. Hence for the confined flows under consideration here, we can find the distribution of piezometric head for a combination of elementary flows by addition.

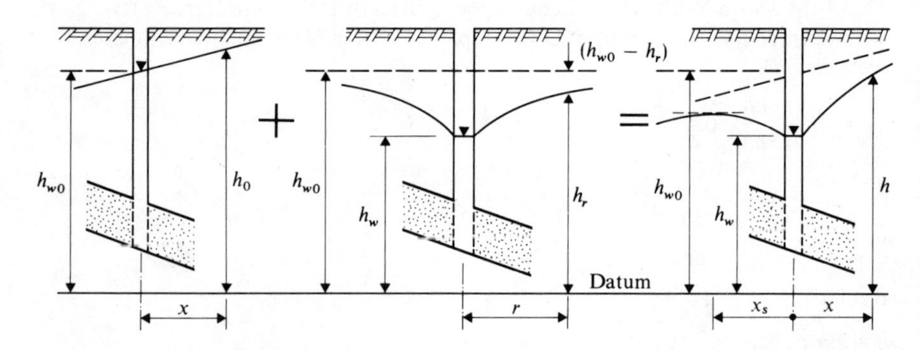

Fig. 3.2. Definition diagram for a well in a confined aquifer of constant thickness carrying uniform flow.

Thus for a well in a uniform flow two elementary flows are superimposed on each other as shown diagramatically in Fig. 3.2. It will be seen that

$$h + (h_{wo} - h_r) = h_0$$

That is,

$$h = h_0 - h_{wo} + h_r \qquad 3.25$$

For the uniform flow, using equation 2.3

$$h_0 = -\frac{U_0}{K}x + h_{wo}$$

and for the well, from equation 3.19 and 3.20

$$h_r = h_w + \frac{q}{2\pi K} \ln \frac{r}{r_w}$$

Hence

$$h = h_w - \frac{U_0}{K} x + \frac{q}{2\pi K} \ln \frac{r}{r_w} \qquad 3.26$$

The potential flow analogy can be used to get the same result by adding the velocity potentials for the two components. Thus, for the uniform flow

$$\phi_0 = -U_0 x \qquad 3.1$$

where U_0 is to be inserted with its sign and for the well

$$\phi_r = \frac{q}{2\pi} \ln \frac{r}{r_w}$$

For the combined flow

$$\phi = \phi_0 + \phi_r = -U_0 x + \frac{q}{2\pi} \ln \frac{r}{r_w} \qquad 3.27$$

and using the equation 3.2

$$Kh + C = -U_0 x + \frac{q}{2\pi} \ln \frac{r}{r_w}$$

The boundary condition $h = h_w$ when $x = 0$ and $r = r_w$ yields

$$Kh_w + C = 0$$

whence

$$h = h_w - \frac{U_0}{K} x + \frac{q}{2\pi K} \ln \frac{r}{r_w} \qquad 3.26$$

$$= h_w - \frac{U_0}{K} x + \frac{q}{4\pi K} \ln \frac{x^2 + y^2}{r_w^2} \qquad 3.28$$

There is no need to evaluate the constant C in equation 3.2 for the individual elementary flows. It is determined by the boundary conditions to be satisfied by h and we are interested in these for the combined flow only. The individual constants and any shift of datum can all be absorbed into C for the combined flow and we have to know what this constant is, but no other.

The discharge to the well is given by

$$q = \frac{2K(h_{w0} - h_w)}{\ln \frac{r_e}{r_w}} \qquad 3.15$$

For graphical superposition, consider first the flow net for the uniform flow. It is a rectangular grid of equipotentials ($x = -\phi_0/U_0$) and streamlines ($y = -\psi_0/U_0$). The contour interval is constant for each function and the flow net of this element is shown in Fig. 3.3.

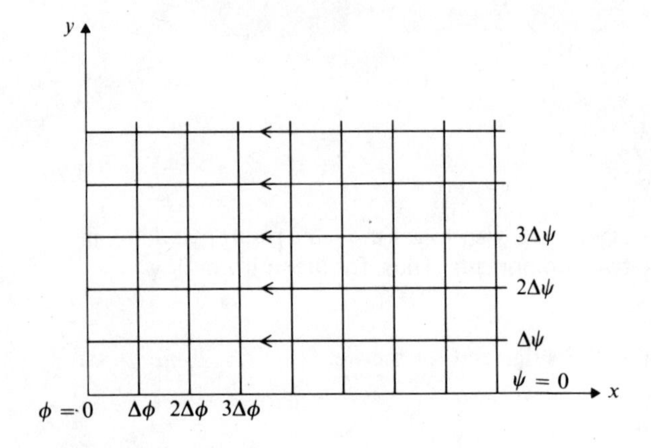

Fig. 3.3. Diagrammatic flow net for uniform flow.

We now want to plot curves for the potential of the well $\phi_r = (q/2\pi) \ln (r/r_w)$, and its stream function $\psi_r = (q/2\pi)\theta$, so that the respective contour intervals will be the same as for the uniform flow. For an equipotential line in the rectangular grid at distance x from the y axis, a corresponding equipotential circle of radius r can be determined using

$$-U_0 x = \phi_0 = \phi_r = \frac{q}{2\pi} \ln \frac{r}{r_w}$$

whence

$$r = r_w \exp \left[-\frac{2\pi U_0}{q} x \right] \qquad 3.29$$

Similarly, for every streamline of the uniform flow at distance y from the x axis, there is a corresponding streamline of the well at an angle θ to $\theta = 0$ given by

$$-U_0 y = \psi_0 = \psi_r = \frac{q}{2\pi} \theta$$

whence

$$\theta = \frac{-2\pi U_0}{q} y \qquad 3.30$$

The flow net for the well is shown in Fig. 3.4

When the two flow nets are superimposed on each other, the flow net for the combined flow may be plotted by drawing contours for $\phi(=\phi_0 + \phi_r)$ and $\psi(=\psi_0 + \psi_r)$ as shown on Fig. 3.5. It will be seen that the resultant streamline pattern separates the flow into two regions; a central region bounded by the streamlines $\psi = \pm\frac{1}{2}q$ and a region outside this curve. The point where the dividing streamline crosses the x axis is a stagnation point, a singularity where the velocity is zero. It will also be recognized as the point where the tangent to

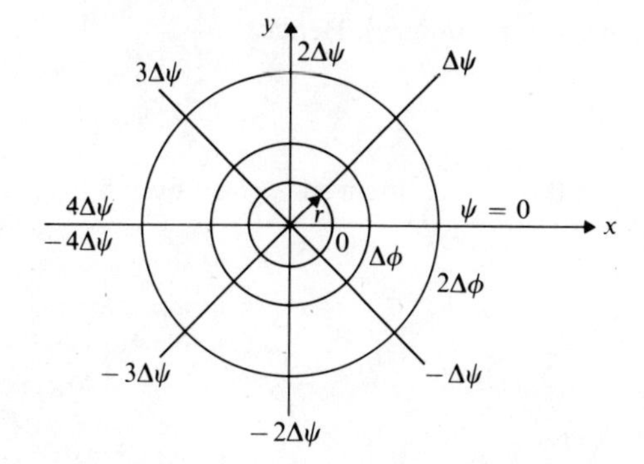

Fig. 3.4. Diagrammatic flow net for radial flow in an aquifer of constant thickness.

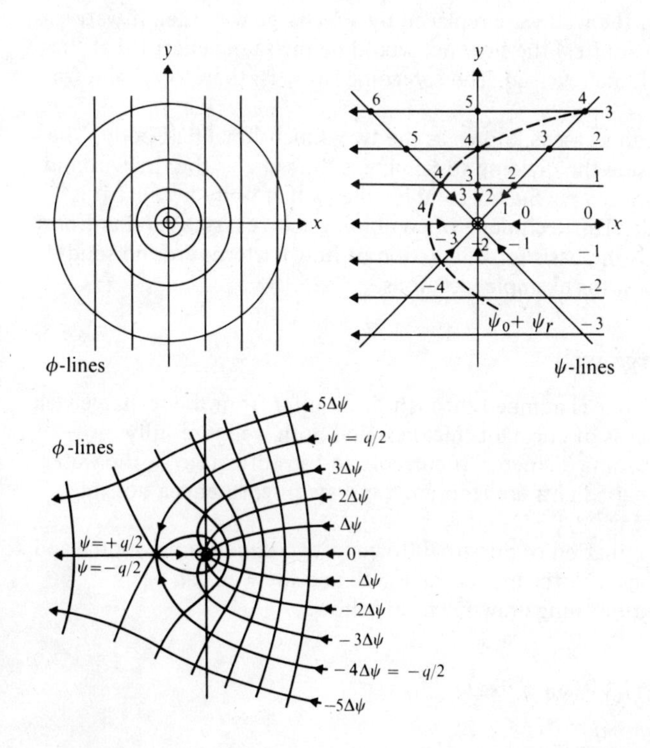

Fig. 3.5. Illustration of the concepts of superposition, uniform flow and radial flow.

the piezometric surface is horizontal (see Fig. 3.2). Its location downstream of the well is found by putting the x component of velocity equal to zero and solving for x

$$u = +U_0 - \frac{q}{2\pi} \frac{x}{x^2 + y^2} = 0$$

At the stagnation point, $x = x_s$ and $y = 0$ (by symmetry). Hence

$$x_s = \frac{q}{2\pi U_0} \tag{3.31}$$

The width of the aquifer contributing to the flow to the well is given by $x = \infty$ and $\psi = \frac{1}{2}q$ yielding for the half width

$$y = \frac{q}{2U_0}$$

and

$$2y = \frac{q}{U_0} \tag{3.32}$$

It is obvious that, if the well were replaced by a recharge well (i.e., if water were pumped into the aquifer) the flow net would be the same, except that the direction of flow would be reversed. The foregoing study is, therefore, valid for a recharge well.

In hydrodynamics, this case is known as the two-dimensional half body. The flow in the region outside the dividing streamline is the same as the flow around a solid body of the same shape. Such a solid is called a half body because it is unbounded on one side. This technique of combining sources, sinks and uniform flows in various ways is the basis for calculation of flow patterns around solid bodies of various shapes, for example, aerofoils.

3.4 Well and recharge well

In this combination, water is pumped through the aquifer from the recharge well to the well. The aquifer is of constant thickness and both wells are fully penetrating and of the same diameter. If currents and stratification of the water caused by temperature gradients are ignored, the system represents a possible industrial cooling system.

The analogous combination of potential flow is a two-dimensional source and a sink. The distributions of h for the recharge well and the well can be superimposed by superimposing drawdowns as follows:
from Fig. 3.6

$$h_{w0} - h = (h_{w0} - h_{r1}) + (h_{w0} - h_{r2})$$

$$= \frac{q}{2\pi K} \ln \frac{r_e}{r_1} - \frac{q}{2\pi K} \ln \frac{r_e}{r_2}$$

$$h = h_{w0} + \frac{q}{2\pi K} \ln \frac{r_1}{r_2} \tag{3.33}$$

Alternatively, the potentials of the well and the recharge well may be added as follows:

$$\phi_{r1} = \frac{q}{2\pi} \ln \frac{r_1}{r_w}$$

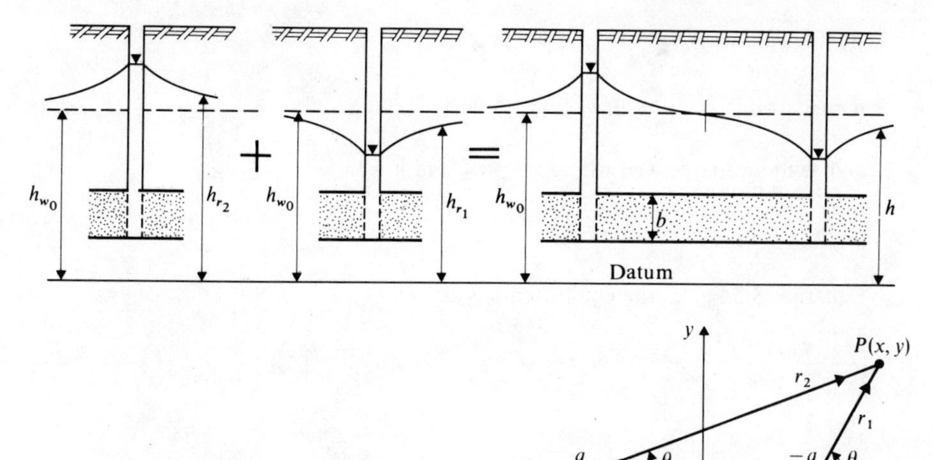

Fig. 3.6. Definition sketch for well and recharge well in an aquifer of constant thickness.

and

$$\phi_{r2} = -\frac{q}{2\pi} \ln \frac{r_2}{r_w}$$

For the two elements together

$$\phi = \phi_{r1} + \phi_{r2}$$

$$= \frac{q}{2\pi} \ln \frac{r_1}{r_2} = \frac{q}{4\pi} \ln \frac{(x - x_1)^2 + y^2}{(x + x_1)^2 + y^2} \qquad 3.34$$

Substituting for ϕ from equation 3.2

$$Kh + C = \frac{q}{2\pi} \ln \frac{r_1}{r_2}$$

The constant C may be found using the condition that $h = h_{wo}$ when $q = 0$, so that

$$C = -Kh_{wo}$$

leading to $\phi = K(h - h_{wo})$ and

$$h = h_{wo} + \frac{q}{2\pi K} \ln \frac{r_1}{r_2}$$

$$= h_{wo} + \frac{q}{4\pi K} \ln \frac{(x - x_1)^2 + y^2}{(x + x_1)^2 + y^2} \qquad 3.35$$

At the well, where $x = x_1$ and $y = \pm r_w$

$$h = h_{w1} = h_{w0} - \frac{q}{2\pi K} \ln \frac{2x_1}{r_w} \qquad 3.36$$

and at the recharge well where $x = -x_1$ and $y = \pm r_w$

$$h = h_{w2} = h_{w0} + \frac{q}{2\pi K} \ln \frac{2x_1}{r_w} \qquad 3.37$$

Equation 3.34 gives the equipotentials as

$$\frac{(x - x_1)^2 + y^2}{(x + x_1)^2 + y^2} = e^{2\pi\phi/q} = A_1 \qquad 3.38$$

i.e.

$$\left(x - \frac{1 + A_1}{1 - A_1} x_1\right)^2 + y^2 = \frac{4A_1 x_1^2}{(1 - A_1)^2} \qquad 3.39$$

Thus, the equipotentials are circles, with centres at

$$x = x_1 \frac{1 + A_1}{1 - A_1}, \quad y = 0 \qquad 3.40$$

and radii given by

$$R_1 = \frac{2\sqrt{A_1}\,x_1}{1 - A_1} \qquad 3.41$$

Along the y axis, $x = 0$ and, from equation 3.38, $A_1 = 1$ and $\phi = 0$. From equation 3.41, $A_1 = 1$ leads to $R = \infty$. Thus, the y axis is seen to be a circle of infinite radius representing the equipotential $\phi = 0$ (corresponding to $h = h_{w0}$). It will also be seen that, if $x = \pm\infty$ and $y \neq \pm\infty$, ϕ is again equal to zero. The well face, of radius r_w with centre at $(x_1, 0)$ is not an equipotential, but it can be assumed to be the equipotential corresponding to $h = h_w$ without serious error. Examination of the equipotential $R_1 = r_w$ will confirm this. It almost coincides with the well face and represents the correct value of h.

Similarly, addition of the stream functions yields

$$\psi = \frac{q}{2\pi}(\theta_1 - \theta_2) = \frac{q}{2\pi}\left[\tan^{-1}\frac{y}{x - x_1} - \tan^{-1}\frac{y}{x + x_1}\right] \qquad 3.42$$

and any particular streamline is defined by

$$\theta_1 - \theta_2 = \frac{2\pi}{q}\psi = \text{constant};$$

$$\frac{\tan\theta_1 - \tan\theta_2}{1 + \tan\theta_1 \tan\theta_2} = \tan\frac{2\pi}{q}\psi = A_2$$

$$\frac{\dfrac{y}{x - x_1} - \dfrac{y}{x + x_1}}{1 + \dfrac{y^2}{(x - x_1)(x + x_1)}} = A_2 \tag{3.43}$$

whence

$$x^2 + \left(y - \frac{x_1}{A_2}\right)^2 = x_1^2 \left(1 + \frac{1}{A_2^2}\right) \tag{3.44}$$

This is another family of circles, with centres on the y axis at $y = x_1/A_2$, the radii being given by

$$R_2 = x_1 \sqrt{1 + \frac{1}{A_2^2}} \tag{3.45}$$

The x axis is a streamline of infinite radius, which can be seen by substituting $y = 0$ in equation 3.43. This yields $A_2 = 0$ and equation 3.45 then gives $R_2 = \infty$. The value of the stream function of the x axis is found from the fact that $A_2 = 0$ gives

$$\theta_1 - \theta_2 = 0 \text{ or } \pm \pi$$

and hence

$$\psi = 0 \text{ or } \pm \frac{q}{2}$$

Reference to Fig. 3.6 will show that $\theta_1 - \theta_2 = \pm \pi$ between $x = x_1$ and $x = -x_1$ so that $\psi = \pm q/2$ on the x axis between the wells. For $x < -x_1$ and $x > x_1$,

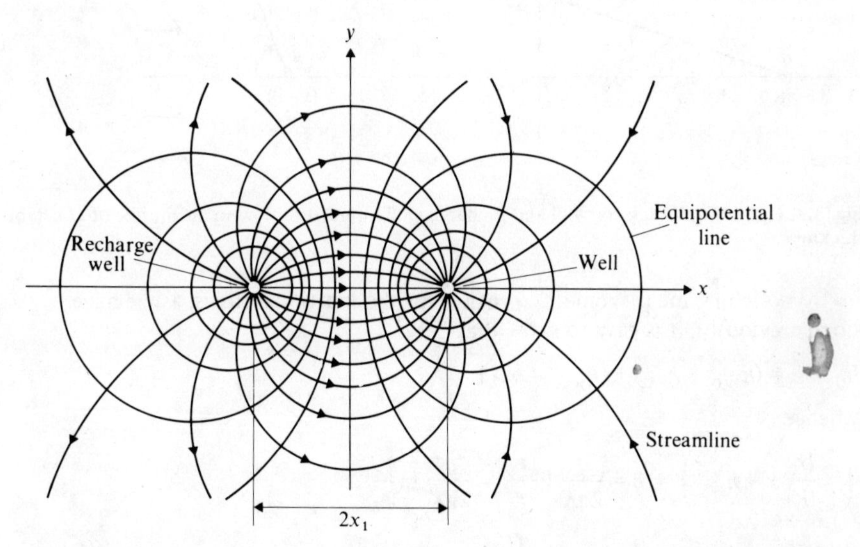

Fig. 3.7. Diagrammatic flow net for well and recharge well flow in an aquifer of constant thickness.

$\theta_1 - \theta_2 = 0$ and there the x axis is the streamline $\psi = 0$. When $\theta_1 - \theta_2 = \pi/2$, $\psi = q/4, A_2 = \infty$ and $R_2 = x_1$. That is, this streamline is a circle of radius x_1 with its centre at the origin of coordinates. The complete flow net for this problem is shown in Fig. 3.7

In such an ideal system in an aquifer extending to infinity, all the water from the recharge well would eventually return to the well, although some of it must make a very long journey. The system is said to have complete recirculation

3.5 Well and recharge well in uniform flow

The analysis above may be extended by placing the sink-source pair in a uniform flow. This system would represent a well and a recharge well in an aquifer carrying a natural flow of groundwater. It is clear that, if the recirculation is to be as small as possible, the recharge well must be downstream of the well. The system to be analysed is defined in Fig. 3.8. The two wells are on the x axis and the uniform flow is in the direction of x decreasing. That is to say, the velocity of the uniform flow, U_0, is negative. Other arrangements are, of course, possible. For example, the recharge well could be placed upstream of the well, or the two wells could be located on the y axis, with the line joining them transverse to the uniform flow. Analysis of these is left to the reader: in principle, they are the same as what follows here.

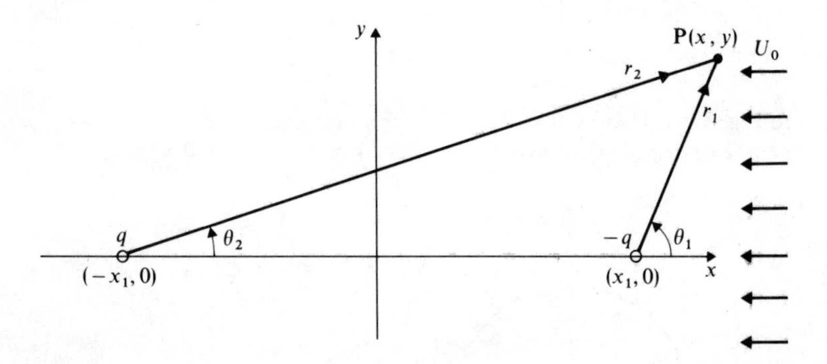

Fig. 3.8. Definition sketch for well and recharge well in uniform flow in an aquifer of constant thickness.

By sketching the piezometric profiles for the individual flows, as we have done previously, it is easy to show that

$$h_0 - h = (h_{w0} - h_{r1}) + (h_{w0} - h_{r2})$$

whence

$$\left(-\frac{U_0}{K}x + h_{w0}\right) - h = \frac{q}{2\pi K}\ln\frac{r_e}{r_1} - \frac{q}{2\pi K}\ln\frac{r_e}{r_2}$$

$$h = h_{w0} - \frac{U_0}{K}x + \frac{q}{2\pi K}\ln\frac{r_1}{r_2} \qquad 3.46$$

$$= h_{w0} - \frac{U_0}{K}x + \frac{q}{4\pi K}\ln\frac{(x - x_1)^2 + y^2}{(x + x_1)^2 + y^2} \qquad 3.47$$

Alternatively, the three potential functions may be added.

$$\phi_0 = -U_0 x$$

$$\phi_{r1} = \frac{q}{2\pi}\ln\frac{r_1}{r_w}$$

$$\phi_{r2} = -\frac{q}{2\pi}\ln\frac{r_2}{r_w}$$

Therefore

$$\phi = \phi_0 + \phi_{r1} + \phi_{r2}$$

$$= -U_0 x + \frac{q}{2\pi}\ln\frac{r_1}{r_2} \qquad 3.48$$

Using equation 3.2 for ϕ

$$Kh + C = -U_0 x + \frac{q}{2\pi}\ln\frac{r_1}{r_2}$$

and if $h = h_{w0}$ at $x = 0$ (where $r_1 = r_2$)

$$Kh_{w0} + C = 0$$

so that

$$C = -Kh_{w0}$$

and

$$\phi = K(h - h_{w0})$$

This value of C also gives

$$h = h_{w0} - \frac{U_0}{K}x + \frac{q}{2\pi K}\ln\frac{r_1}{r_2} \qquad 3.46$$

By superimposing the stream functions,

$$\psi = -U_0 y + \frac{q}{2\pi}(\theta_1 - \theta_2)$$

$$= -U_0 y + \frac{q}{2\pi}\left(\tan^{-1}\frac{y}{x - x_1} - \tan^{-1}\frac{y}{x + x_1}\right) \qquad 3.49$$

The streamline pattern which emerges as a result of this superposition will belong to one of the three possible types shown in Fig. 3.9, depending on the magnitudes of q, x_1 and U_0.

In case (a) there is no recirculation and each well has its own boundary streamline. The points 1 and 2 are stagnation points, given by $y = 0$, $x = x_s$ and $u = 0$.

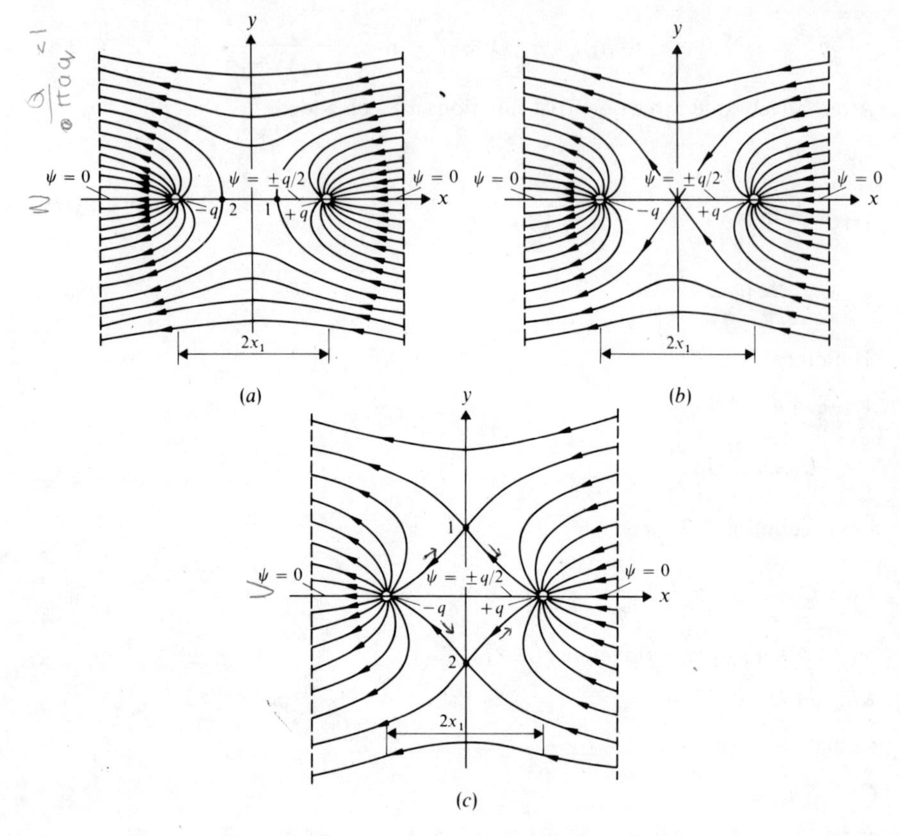

Fig. 3.9. Diagrammatic flow nets for well and recharge well in uniform flow in an aquifer with constant thickness. (*a*) No recirculation, two stagnation points on *x* axis. (*b*) No recirculation, one stagnation point on *y* axis. (*c*) Recirculation, two stagnation points on *y* axis.

$$u = -\frac{\partial \phi}{\partial x} = U_0 - \frac{q}{2\pi}\left[\frac{x - x_1}{(x - x_1)^2 + y^2} - \frac{x + x_1}{(x + x_1)^2 + y^2}\right] \tag{3.50}$$

Substitution of the appropriate values of x, y and u gives

$$x_s = \pm x_1 \sqrt{1 + \frac{q}{\pi x_1 U_0}} \tag{3.51}$$

In case (b) the two stagnation points merge into one which coincides with the origin of coordinates. That is to say $x_s = 0$, whence

$$-\frac{q}{\pi x_1 U_0} = 1 \tag{3.52}$$

in which U_0 has to be inserted with its proper sign.

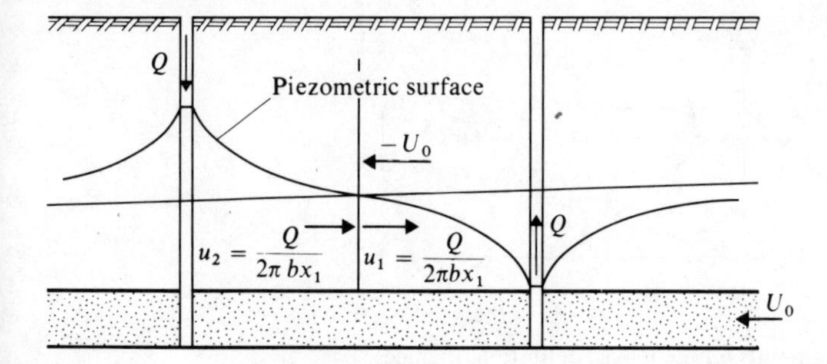

Fig. 3.10. Elevation along the x-axis for case (b) of Fig. 3.9.

This result could also be obtained by superimposing the velocity of flow from the recharge well on the velocity to the well at the origin and putting this equal to the uniform flow velocity, Fig. 3.10.

The condition for stagnation at the origin is

$$-U_0 = u_1 + u_2 = \frac{q}{\pi x_1}$$

i.e.

$$-\frac{q}{\pi x_1 U_0} = 1 \tag{3.52}$$

Here we have applied the principle of superposition to a component of velocity.

In case (c) there are again two stagnation points, this time on the y axis. Their coordinates are given by $x = 0, y = y_s$ and $u = 0$. Substitution of these in equation 3.50 gives

$$y_s = \pm x_1 \sqrt{-\frac{q}{\pi x_1 U_0} - 1} \tag{3.53}$$

In this case, some water flows from the recharge well to the well, and we want to know how much. We get this information from the stream function.

For any point on the y axis (where $x = 0$)

$$\psi_{x=0} = -U_0 y + \frac{q}{2\pi}(\theta_1 - \theta_2)$$

$$= -U_0 y + \frac{q}{2\pi} 2\left[\frac{\pi}{2} - \tan^{-1}\frac{y}{x_1}\right]$$

$$= \frac{q}{2} - U_0 y - \frac{q}{\pi}\tan^{-1}\frac{y}{x_1}$$

as may be seen in Fig. 3.11.

At the stagnation point where $y = +|y_s|$

$$\psi_s = \frac{q}{2} - U_0\,|y_s| - \frac{q}{\pi}\tan^{-1}\frac{|y_s|}{x_1}$$

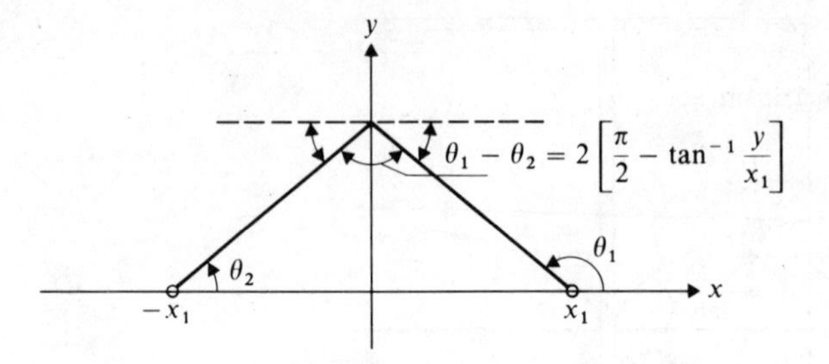

Fig. 3.11. Stagnation points on the y-axis, definition of angles.

and at the origin (substituting $y = 0$)

$$\psi_{0,\,0} = \frac{q}{2}$$

The total discharge recirculating is therefore given by

$$q_r = 2(\psi_{0,\,0} - \psi_s)$$

$$= 2\frac{q}{\pi}\tan^{-1}\frac{|y_s|}{x_1} + 2U_0\,|y_s| \qquad 3.54$$

Hence

$$\frac{q_r}{q} = \frac{2}{\pi}\left\{\tan^{-1}\sqrt{-\frac{q}{\pi U_0 x_1} - 1} + \frac{\sqrt{-\frac{q}{\pi U_0 x_1} - 1}}{\frac{q}{\pi U_0 x_1}}\right\} \qquad 3.55$$

Thus, recirculation will or will not occur, depending on whether $-(q/\pi U_0 x_1) \gtrless 1$ and the proportion of the total flow injected which will be recirculated is determined by the same quantity.

3.6 Well and lake

The flow nets mapped in the preceding section can be readily adapted to describe other flows. This is possible if an equipotential or a streamline in the flow net can be made to represent a boundary in the actual groundwater system where the piezometric head or the stream function is constant. For example, the groundwater system may be as sketched in Fig. 3.12 (i).

The confined aquifer ends in a vertical plane on the shore of a lake. The piezometric head is constant over this plane, which appears as a straight line when plotted on the x-y plane. Thus, in the flow net on Fig. 3.7 the y-axis could represent the shoreline; it is an equipotential and it is straight. If the flow to the well is all from the lake, we can say that the y-axis is a plane source in the equivalent potential flow system. The real well is equivalent to a sink as before and the recharge well or source is imaginary. Physically it does not exist and it is

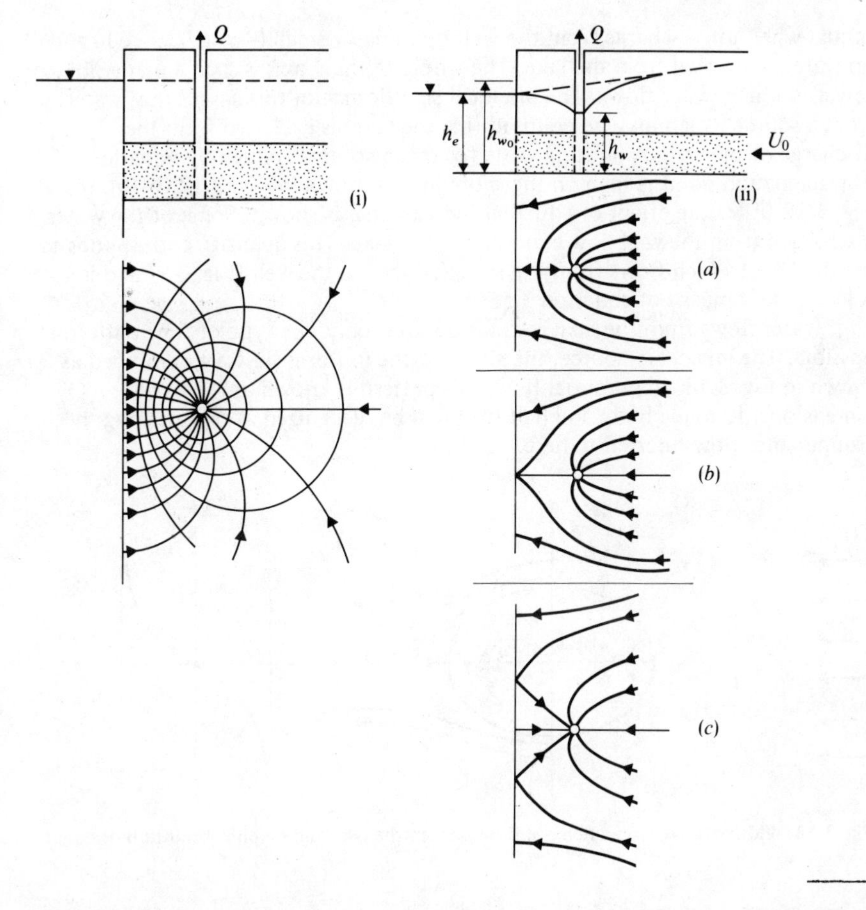

Fig. 3.12. Diagrammatic flow nets for a well in an aquifer of constant thickness with uniform flow near a lake.

introduced solely as a convenient device to enable the correct boundary condition to be satisfied at the shoreline. It will also be seen that if there is a flow in the aquifer towards the lake (Fig. 3.12 (ii)), part or all of the y-axis will be a plane sink and only part of it may be a plane source. The flow net will then be the same as the right side of Fig. 3.9. It is also apparent that a curved shoreline which can be represented by a circular arc can be dealt with in the same way.

These elementary examples illustrate the method of images in which imaginary image sources or sinks are used. This device plays a very important role when complex variable and conformal transformation methods are used to solve more complicated problems.

When there is a flow through the aquifer towards the lake as shown in Fig. 3.12 (ii), three flow nets are possible, corresponding to the three possibilities discussed in the preceding section. The flow pattern shown in Fig. 3.12 (ii)(*a*) occurs when the discharge from the well is relatively small ($-q/\pi U_0 x_1 < 1$) and no water is diverted from the lake. The whole of the y axis is then a plane sink, towards which water flows. The practical significance of this case is that water from a source which may be polluted (i.e. the lake) is excluded from the discharge of the well. A limiting value for this discharge is given by equation 3.52. The stagnation point is then on the shoreline, as shown in Fig. 3.12 (ii) (*b*). In Fig. 3.12 (ii)(*c*) the effect of a further increase in q is shown. Some of the water discharged from the well now comes from the lake. This quantity corresponds to the discharge which flows from the recharge well to the well (Fig. 3.9) and is calculated by means of equation 3.55.

If water flows from the lake into the aquifer, only one type of flow pattern is possible. The imaginary source, the sink and the uniform flow are combined as shown in Fig. 3.13. In potential flow, this pattern is known as the two-dimensional Rankine body and half of it will be recognized as representing the groundwater flow referred to here.

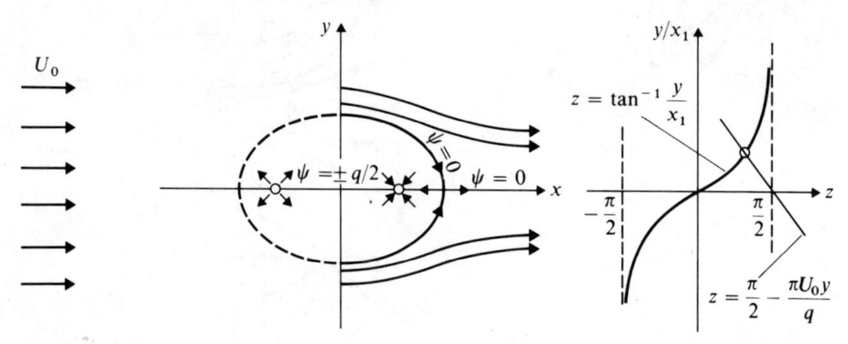

Fig. 3.13. Width of diversion when the flow is from the lake and graphical solution of equation 3.59.

The stream function is given by

$$\psi = -U_0 y + \frac{q}{2\pi}(\theta_1 - \theta_2)$$

$$= -U_0 y + \frac{q}{2\pi}\left[\tan^{-1}\frac{y}{x - x_1} - \tan^{-1}\frac{y}{x + x_1}\right] \qquad 3.56$$

Therefore

$$u = -\frac{\partial \psi}{\partial y}$$

$$= U_0 - \frac{q}{2\pi}\left[\frac{x - x_1}{(x - x_1)^2 + y^2} - \frac{x + x_1}{(x + x_1)^2 + y^2}\right] \qquad 3.57$$

At the stagnation point, $y = 0$ (from symmetry) $x = x_s$ and $u = 0$. Hence

$$x_s = \pm x_1 \sqrt{1 + \frac{q}{\pi U_0 x_1}} \qquad 3.58$$

Only the positive value has any meaning, since none of the flow pattern on the left of the y axis is real. From equation 3.56 it will be seen that the x axis between the shoreline and the well is the streamline

$$\psi = \pm\frac{q}{2} \qquad (y = 0, \theta_1 = \pm\pi, \theta_2 = 0)$$

Beyond the well it is the streamline $\psi = 0$. It follows that, since the stagnation point is on this part of the x axis, the dividing streamline is also given by $\psi = 0$. The length of the shoreline is given by substituting $x = 0$ in the equation $\psi = 0$. Thus, from equation 3.57

$$-U_0 y + \frac{q}{2\pi}(\theta_1 - \theta_2) = 0$$

$$-U_0 y + \frac{q}{2\pi}\left(\pi - 2\tan^{-1}\frac{y}{x_1}\right) = 0$$

$$\tan\left(1 - \frac{2U_0 y}{q}\right)\frac{\pi}{2} = \frac{y}{x_1} \tag{3.59}$$

When solved for y, equation 3.59 gives the half width of the diversion (Fig. 3.13),

$$\tan^{-1}\frac{y}{x_1} = \frac{\pi}{2} - \frac{\pi U_0 y}{q} \equiv z;\ z = \tan^{-1}\left(\frac{y}{x_1}\right)$$

The groundwater potential is given by

$$\phi = -U_0 x + \frac{q}{2\pi}\ln\frac{r_1}{r_2} \tag{3.60}$$

whence, using equation 3.2

$$Kh + C = -U_0 x + \frac{q}{2\pi}\ln\frac{r_1}{r_2} \tag{3.61}$$

Since $r_1 = r_2$ and $h = h_e$ when $x = 0$

$$Kh_e = C = 0$$

$$C = -Kh_e$$

and

$$h = h_e - \frac{U_0}{K}x + \frac{q}{2\pi K}\ln\frac{r_1}{r_2} \tag{3.62}$$

The discharge to the well may be found from equation 3.62 by noting that $h = h_w$ at the face of the well where $x = x_1, r_1 = r_w$ and $r_2 \doteqdot 2x_1$

$$h_w = h_e - \frac{U_0}{K}x_1 + \frac{q}{2\pi K}\ln\frac{r_w}{2x_1} \tag{3.63}$$

$$q = \frac{2\pi K\left[\left(h_e - \frac{U_0}{K}x_1\right) - h_w\right]}{\ln \frac{2x_1}{r_w}}$$

$$= \frac{2\pi K(h_{wo} - h_w)}{\ln \frac{2x_1}{r_w}} \qquad 3.64$$

Here the substitution $h_{wo} = h_e - U_0/K\, x_1$ (see Fig. 3.12 (ii)) has been made. Note that for the same drawdown, addition of a uniform flow alters the distribution of flow to the well, but not its rate, i.e. the strength of the sink and the source remains the same. It is interesting to note that the discharge to the well is the same as that for flow to a well when the external radius r_e (where $h = h_e$) equals $2x_1$. Substitution from equation 3.64 into equation 3.62 gives, for the piezometric head,

$$h = \left(h_e - \frac{U_0}{K}x\right) + \left[\left(h_e - \frac{U_0}{K}x_1\right) - h_w\right] \frac{\ln \frac{r_1}{r_2}}{\ln \frac{2x_1}{r_w}}$$

Lastly, it should be noted that, if Fig. 3.13 is rotated so that the y axis is horizontal, the lower half of the diagram represents the flow pattern for seepage to a horizontal drain from a flooded area above it. In this case, the surface of the flooded ground is an equipotential. By adding the potential and stream function, seepage into a large number of parallel drains might be investigated. However, without more sophisticated analytical tools than those used here, the work is quite tedious.

3.7 Eccentrically located well; multi-layer aquifer

The combination of a well and an imaginary recharge well can be used to find the flow net to a well located off-centre in a circular island in a reservoir (c.f. radial flow to a well). The problem is defined in Fig. 3.14 and it will be seen that we have to decide where to put the image well (i.e. to calculate x_1) and to make sure that one of the circular equipotentials coincides with the shoreline of the island. Using equation 3.40 and 3.41, which give the position of the centre and the radius of any equipotential, we have

$$x_1 + x_0 = \frac{1 + A_1}{1 - A_1}x_1 \qquad 3.65$$

and

$$r_e = \frac{2\sqrt{A_1}}{1 - A_1}x_1 \qquad 3.66$$

Solving for x_1 and A_1, we get

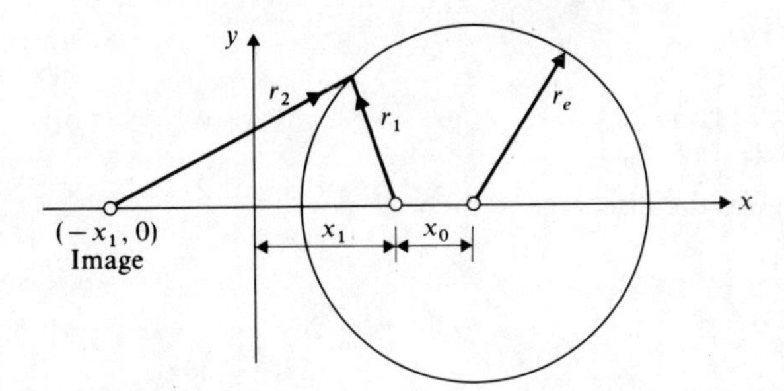

Fig. 3.14. Definition sketch for an eccentrically located well in an aquifer of constant thickness.

$$x_1 = \frac{1 - \left(\frac{x_0}{r_e}\right)^2}{2\frac{x_0}{r_e}} r_e \qquad 3.67$$

and

$$A_1 = \left(\frac{x_0}{r_e}\right)^2 \qquad 3.68$$

Equation 3.67 gives the location of the image well immediately and equation 3.68 enables us to calculate the distribution of h as follows:

$$Kh + C = \phi = \frac{q}{2\pi} \ln \frac{r_1}{r_2}$$

On the shoreline of the island, $h = h_e$ and

$$\phi = \frac{q}{4\pi} \ln A_1 = \frac{q}{2\pi} \ln \frac{x_0}{r_e}$$

whence

$$C = \frac{q}{2\pi} \ln \frac{x_0}{r_e} - Kh_e$$

and

$$h_e - h = \frac{q}{2\pi K} \ln \frac{x_0}{r_e} \frac{r_2}{r_1} \qquad 3.69$$

The drawdown at the well is given by substituting in equation 3.69 $h = h_w$, $r_1 = r_w$ and

$$r_2 = 2x_1 = \frac{1 - \left(\frac{x_0}{r_e}\right)^2}{\frac{x_0}{r_e}} r_e$$

yielding

$$h_e - h_w = \frac{q}{2\pi K} \ln \left\{ \left(1 - \left[\frac{x_0}{r_e} \right]^2 \right) \frac{r_e}{r_w} \right\} \qquad 3.70$$

Rearranging equation 3.70 gives the discharge to the well

$$q = \frac{2\pi K \, (h_e - h_w)}{\ln \left\{ \left(1 - \left[\frac{x_0}{r_e} \right]^2 \right) \frac{r_e}{r_w} \right\}} \qquad 3.71$$

Equations 3.70 and 3.15 give the ratio of discharges for the eccentrically and centrally located wells for equal drawdown as

$$\frac{q_c}{q_e} = \frac{\ln \{[1 - (x_0/r_e)^2] \, r_e/r_w\}}{\ln r_e/r_w}$$

or for equal discharges the ratio of the drawdowns, which is the reciprocal of the above ratio.

In all of the foregoing examples, flow in a single confined aquifer is analysed. However, in nature wells frequently penetrate two or more aquifers and flow to such a well is of practical interest. Studies of the multi-aquifer problem have been been undertaken by Sokol (1963), Papadopulos (1966) and Abdul Khader *et al.* (1973).

For any aquifer

$$\frac{\partial^2 s_n}{\partial r^2} + \frac{1}{r} \frac{\partial s_n}{\partial r} = \frac{S}{T} \frac{\partial s_n}{\partial t}$$

$$s(r, 0) = 0$$

$$s(\infty, t) = 0$$

$$2\pi T_n r \frac{\partial s_n}{\partial r} = - Q_n(t) \text{ as } r \to 0$$

At the well

$$\Sigma Q_i = 0 \text{ for } t \leq 0$$

$$= Q \text{ for } t > 0$$

$$h_1 - s_1 = h_2 - s_2 = \ldots = h_n - s_n = \ldots |_{r=r_w}$$

Such a set of equations cannot be solved exactly, but a technique for numerical solution is discussed by Sternberg (1969, 1973).

3.8 Horizontal drains in porous media

The reader can easily verify that superposition of two wells leads to the streamline pattern shown in Fig. 3.16.

Both axes are straight streamlines and the origin is a stagnation point. Parts of this pattern can be used for flows with straight boundaries which must be

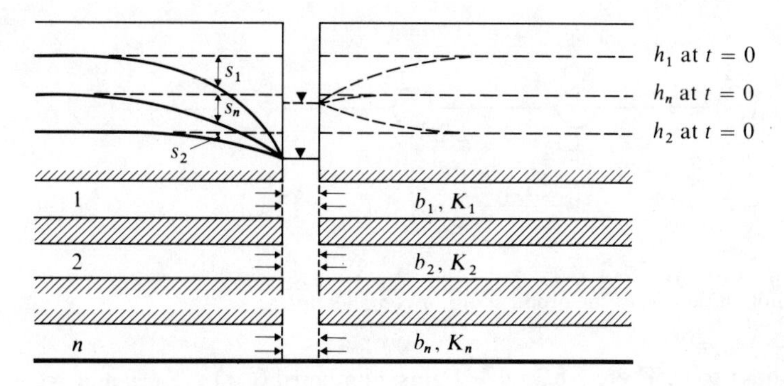

Fig. 3.15. Multi-layer aquifer.

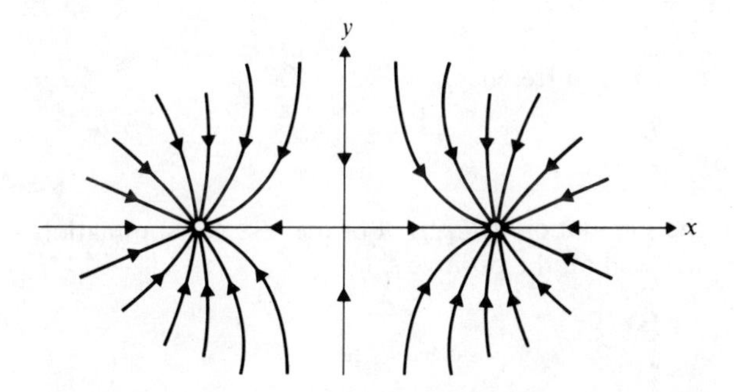

Fig. 3.16. Diagrammatic streamline pattern for two wells in an aquifer of constant thickness.

streamlines. For example, the y axis can represent a plane horizontal impervious boundary. The streamlines for either half of Fig. 3.16 would then represent flow through a semi-infinite porous medium to a drain a distance x_1 above or below the boundary. Here, the method of images is used again.

Alternatively, the x axis could represent a similar boundary. Half the pattern would then represent flow to a pair of drains half embedded in a confining layer above or below a semi-infinite porous medium. In the simplest case of all, there is only one well. Since the medium is large, the velocity of flow far from the drain will be so small that the pressure distribution beyond some large radius will be hydrostatic. That is, we can assume there is a semicircular equipotential at this radius. The potential is then given by

$$\phi = \frac{q}{\pi} \ln \frac{r}{r_0}$$

where r_0 is the radius of the drain.

Flow to a large number of drains half embedded in the same way can be investigated by superimposing potentials as required. The system of drains is shown in Fig. 3.17 and we consider the potential difference between a point A on the periphery of the centre drain and a point B half way between this drain

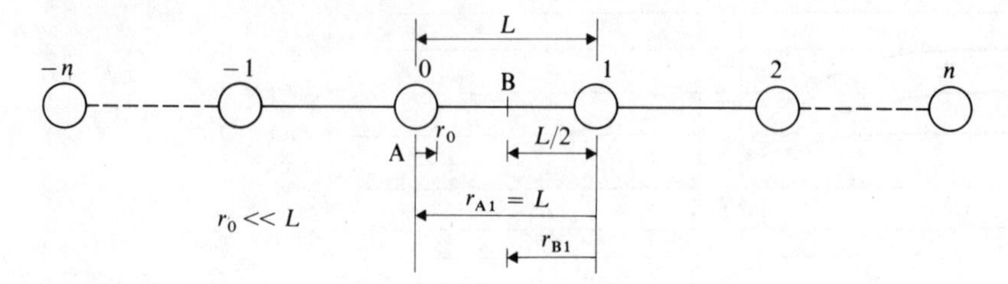

Fig. 3.17. Definition sketch for the problem of horizontal drains.

and the one next to it. There are $2n + 1$ drains, numbered $0, +1, \ldots +n$ and $-1, \ldots -n$ and the radius of each is r_0. The required potential drop is given by

$$\Delta\phi = \phi_B - \phi_A = \frac{q}{\pi}\left\{\sum_{i=0}^{i=+n} \ln \frac{r_{Bi}}{r_{Ai}}\right\} \tag{3.72}$$

Now for the centre well, or the first difference,

$$r_{A0} = r_0 \qquad \text{and} \qquad r_{B0} = \frac{L}{2}$$

giving for the first term of the summation $\ln L/2r_0$. For the rest, the summation is symmetrical i.e. for the first well on the positive side

$$r_{A1} = L \qquad \text{and} \qquad r_{B1} = \frac{L}{2}$$

and so on, until, for the ith well on the positive side

$$r_{Ai} = iL \qquad \text{and} \qquad r_{Bi} = \frac{2i-1}{2}L$$

On the negative side, for the ith well

$$r_{A(-i)} = iL \qquad \text{and} \qquad r_{B(-i)} = \frac{2i+1}{2}L$$

Hence,

$$\Delta\phi = \frac{q}{\pi}\left\{\ln \frac{L}{2r_0} + \sum_{i=\pm 1}^{i=\pm n} \ln \frac{r_{Bi}}{r_{Ai}}\right\}$$

If we now take pairs of corresponding terms from the two summations and add these

$$\Delta\phi = \frac{q}{\pi}\left\{\ln \frac{L}{2r_0} + \sum_{i=1}^{i=n} \ln\left(\frac{2i-1}{2i}\,\frac{2i+1}{2i}\right)\right\}$$

$$= \frac{q}{\pi}\left\{\ln \frac{L}{2r_0} + \sum_{i=1}^{i=n} \ln\left(1 - \frac{1}{4i^2}\right)\right\}$$

$$= \frac{q}{\pi}\left\{\ln \frac{L}{2r_0} + \ln\left[\prod_{i=1}^{i=n}\left(1 - \frac{1}{4i^2}\right)\right]\right\} \tag{3.73}$$

If n is infinite,

$$\prod_{i=1}^{i=n}\left(1-\frac{1}{4i^2}\right)=\frac{2}{\pi}\sin\frac{\pi}{2}=\frac{2}{\pi}$$

and

$$\Delta\phi=\frac{q}{\pi}\left(\ln\frac{L}{2r_0}-0.454\right) \tag{3.74}$$

Equation 3.74 can be used to describe the case of a large number of drains in an unconfined homogeneous soil which receives a steady rainfall, if the permeability of the soil is large relative to the rate of rainfall so that the rise of the water table between the drains is small enough to be neglected. In the analysis it is assumed that an even number m of line sources of strength q/m are uniformly distributed between each pair of drains (van Schilfgaarde *et al.* (1956)). By the same reasoning as above, the potential drop from B to A is

$$\Delta\phi=\frac{q}{\pi}\left\{\ln\frac{r_{B0}}{r_{A0}}+\sum_{i=1}^{i=\pm n}\ln\frac{r_{Bi}}{r_{Ai}}\right\}$$

$$+\sum_{j=1}^{j=m}\frac{q}{\pi j}\left(\sum_{i=0}^{i=\pm n}\ln\frac{r_{Aji}}{r_{Bji}}\right) \tag{3.75}$$

where r_{Aji} is the distance from A to the jth source to the right of the ith drain.

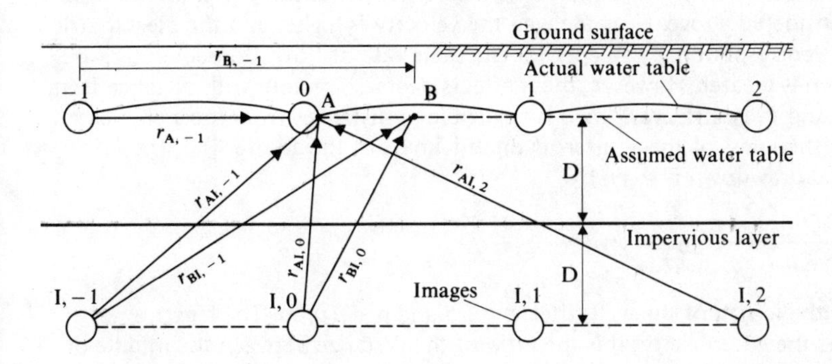

Fig. 3.18. Definition sketch for the problem of horizontal drains in a soil of finite depth.

The authors show that the sum over j vanishes and equation 3.75 reduces to equation 3.74. The analysis can be extended further to describe drainage of a soil of finite depth as shown on Fig. 3.18. By summing all the effects, the potential difference becomes

$$\Delta\phi=\frac{q}{\pi}\left\{\sum_{i=0}^{i=\pm n}\ln\frac{r_{Bi}}{r_{Ai}}+\sum_{i=0}^{i=\pm n}\ln\frac{r_{BIi}}{r_{AIi}}\right\} \tag{3.76}$$

Summation of the additional two terms is laborious. The result is (Hooghoudt,

1940), for $n = \infty$,

$$\Delta\phi = \frac{q}{\pi}\left\{\ln\frac{L}{2r_0} - 0.454 + \tfrac{1}{2}\sum_{1}^{\infty}\ln\frac{\left[(2i-1)^2\frac{L^2}{4} + 4D^2\right]^2}{(i^2L^2 + 4D^2)[(n-1)^2L^2 + 4D^2]}\right\} \tag{3.77}$$

3.9 Remarks

In the preceding section, a number of elementary examples have been discussed. They are elementary in four respects: the wells have been assumed to be fully penetrating, the boundary conditions were simple, the permeability and the thickness of the aquifer were both constant and only simple mathematical techniques were used. In practical problems the wells might not penetrate the aquifer fully and several sources and sinks might be grouped inside a boundary of complicated shape. Furthermore, the fluid properties and the permeability might both vary from place to place. Such complications demand the use of more powerful analytical tools. However, the logic introduced in these simple examples is also basic to the solution of the more complex problems. To deal with these, a working knowledge of complex variable and conformal transformation techniques is essential.

When the thickness of the aquifer is large full penetration is not justified because of cost. The pattern of flow to a partially penetrating well differs from those considered above. Near the well the velocity is higher and the piezometric surface is correspondingly steeper for the same rate of flow. Consequently, the drawdown is greater. However, these effects decrease rapidly with distance from the well and the extra drawdown is negligible at distances from the well greater than the thickness of the aquifer. If the thickness of the aquifer is b, the additional drawdown is given by

$$\Delta s_w = \frac{Q}{2\pi Kb}\,\frac{1-p}{p}\,\ln\frac{\alpha l}{r_w}$$

Here l is the length of the well filter, $p = l/b$ and $\alpha = f(p, e)$. The length e is known as the eccentricity; it is the ratio of the distance between the middle of the filter and the middle of the aquifer to the thickness of the aquifer. The following table is from the Netherlands Hydrological Colloquim (1964):

	$e = 0$	0.05	0.10	0.15	0.20	0.25	0.30	0.35	0.40	0.45
$p = 0.1$	$\alpha = 0.54$	0.54	0.55	0.55	0.56	0.57	0.59	0.61	0.67	1.09
0.2	0.44	0.44	0.45	0.46	0.47	0.49	0.52	0.59	0.89	
0.3	0.37	0.37	0.38	0.39	0.41	0.43	0.50	0.74		
0.4	0.31	0.31	0.32	0.34	0.36	0.42	0.62			
0.5	0.25	0.26	0.27	0.29	0.34	0.51				
0.6	0.21	0.21	0.23	0.27	0.41					
0.7	0.16	0.17	0.20	0.32						
0.8	0.11	0.13	0.22							
0.9	0.06	0.12								

Even with these simple examples, it must be kept in mind that the solutions are valid only provided Darcy's law is valid, a point that should always be checked by calculation of the Reynolds number. Also, the water level in a real well is not h_w because of the screen loss which has to be added to the water level. Frequently clogging of the screens has to be allowed for. The situation where Darcy's law fails near the well is quite common in practice and calls for the two-regime flow solution. In the two-regime solution $dh/dl = a\,|q| + b\,|q|^n$, or some similar expression (c.f. Section 1.8), where a, b are constants and n is a function of the distance from the well. The treatment of this problem is beyond the scope of this introductory text and the reader is referred to the studies by Dudgeon *et al.* (1972, 1973), Huyakorn (1973) and Huyakorn and Dudgeon (1974). Huyakorn and Dudgeon (1972) also prepared an annotated bibliography of groundwater and well hydraulics.

Example 3.1

(a) A fully penetrating well is constructed to pump water from a confined aquifer in which the undisturbed velocity is U_0 parallel to the axis $0x$ and in the direction of x increasing. Sketch curves showing how the x component of velocity is distributed along lines parallel to $0y$ at distances x_s, $2x_s$ and $3x_s$ downstream of the well.
(b) If the thickness of the aquifer is 8 m, its permeability K is 2.5×10^{-4} m/s, the undisturbed gradient of piezometric head is 1 in 1200 and water is pumped from the well at 500 1/hr, calculate the distance of the stagnation point from the well and the width of the aquifer supplying water to the well.

(a) From equation 3.28

$$-K\frac{\partial h}{\partial x} = U_0 - \frac{q}{2\pi}\frac{x}{x^2 + y^2}$$

Let the x-component of velocity at $x = nx_s$ be U_n

$$U_n = -K\left.\frac{\partial h}{\partial x}\right|_{x=nx_s}$$

$$= U_0 - \frac{q}{2\pi}\frac{nx_s}{n^2x_s^2 + y^2}$$

$$= U_0 - \frac{q}{2\pi x_s}\frac{n}{n^2 + (y/x_s)^2}$$

From equation 3.31

$$\frac{q}{2\pi x_s} = U_0$$

Hence,

$$\frac{U_n}{U_0} = 1 - \frac{n}{n^2 + (y/x_s)^2}$$

y/x_s = 0	0.5	1.0	1.5	2.0	2.5	3.0	3.5	4.0
$1^2 + (y/x_s)^2$ = 1	1.25	2	3.25	5	7.25	10	13.25	17
$2^2 + (y/x_s)^2$ = 4	4.25	5	6.25	8	10.25	13	16.25	20
$3^2 + (y/x_s)^2$ = 9	9.25	10	11.25	13	15.25	18	21.25	25
U_1/U_0 = 0	0.2	0.5	0.692	0.8	0.862	0.9	0.925	0.941
U_2/U_0 = 0.5	0.529	0.6	0.68	0.75	0.805	0.846	0.877	0.9
U_3/U_0 = 0.656	0.676	0.7	0.734	0.77	0.803	0.834	0.859	0.88

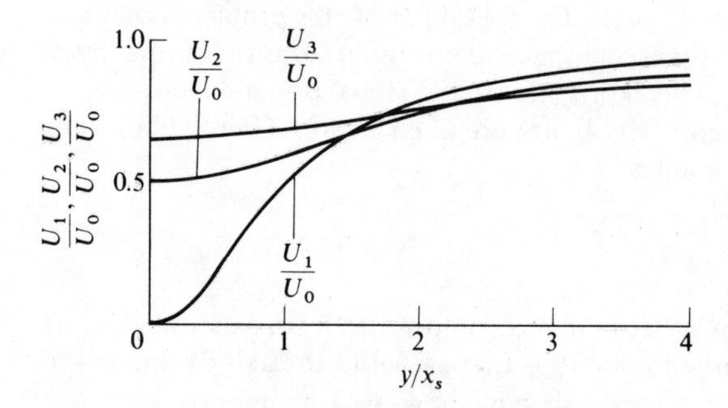

Fig. 3.19. Distribution of discharge velocity downstream of a well.

These curves are sketched in Fig. 3.19.

(b) $U_0 = -2.5 \times 10^{-4} \times (-1/1200) = 2.08 \times 10^{-7}$ m/s

$$q = \frac{500 \times 10^{-3}}{3600} \times \frac{1}{8} = 1.74 \times 10^{-5} \text{m}^2/\text{s}$$

From equation 3.31

$$x_s = \frac{q}{2\pi U_0} = \frac{1.74 \times 10^{-5}}{2 \times \pi \times 2.08 \times 10^{-7}} = 13.3 \text{ m}$$

From equation 3.32

$$\text{width} = \frac{q}{U_0} = \frac{1.74 \times 10^{-5}}{2.08 \times 10^{-7}} = 83.7 \text{ m}$$

Example 3.2

(a) Water flows in a confined aquifer towards a lake with a straight shoreline. The velocity in the aquifer is U_0 in the direction of x decreasing (i.e. U_0 is negative). A fully penetrating well is installed to pump water from the aquifer. Sketch curves showing the distribution along the shoreline of the velocity in the aquifer if (i) $x_s = 0$, (ii) $x_s = 1/2\, x_1$, (iii) $x_s = 2/3\, x_1$ (see Figs. 3.10, 3.12).
(b) If the thickness of the aquifer is 12 m, $U_0 = -3.5 \times 10^{-8}$ m/s and $x_1 = 30$ m, what is the maximum rate at which water can be pumped from the well if lake water is to be excluded?

(a) From equation 3.50

$$u = U_0 - \frac{q}{2\pi}\left[\frac{x - x_1}{(x - x_1)^2 + y^2} - \frac{x + x_1}{(x + x_1)^2 + y^2}\right]$$

$$\frac{u}{U_0} = 1 - \frac{q}{2\pi x_1 U_0}\left[\frac{\frac{x}{x_1} - 1}{\left(\frac{x}{x_1} - 1\right)^2 + \left(\frac{y}{x_1}\right)^2} - \frac{\frac{x}{x_1} + 1}{\left(\frac{x}{x_1} + 1\right)^2 + \left(\frac{y}{x_1}\right)^2}\right]$$

At the shoreline $x = 0$

$$\frac{u}{U_0} = 1 + \frac{q}{\pi x_1 U_0}\left(\frac{1}{1 + (y/x_1)^2}\right)$$

Let $u = U_1$ for $x_s = 0$, case (i)

$u = U_2$ for $x_s = \frac{1}{2}x_1$, case (ii)

$u = U_3$ for $x_s = \frac{2}{3}x_1$, case (iii)

(i) From equation 3.52

$$-\frac{q}{\pi x_1 U_0} = 1$$

Hence,

$$\frac{U_1}{U_0} = 1 - \frac{1}{1 + \left(\frac{y}{x_1}\right)^2}$$

(ii) From equation 3.51, with $x_s = 1/2\, x_1$

$$-\frac{q}{\pi x_1 U_0} = \frac{3}{4}$$

Hence,

$$\frac{U_2}{U_0} = 1 - \frac{3}{4}\,\frac{1}{1 + \left(\frac{y}{x_1}\right)^2}$$

(iii) From equation 3.51, with $x_s = 2/3\, x_1$

$$-\frac{q}{\pi x_1 U_0} = \frac{5}{9}$$

Hence,

$$\frac{U_3}{U_0} = 1 - \frac{5}{9}\,\frac{1}{1 + \left(\dfrac{y}{x_1}\right)^2}$$

To facilitate comparison with the results of Example 3.1, it is better to use the length x_s' instead of x_1 as a scale for y, this length being the distance from the well to its stagnation point i.e. $x_s' = x_1 - x_s$.

When $x_s = 0$, $\qquad x_1 = x_s'$

$$x_s = \frac{1}{2}x_1, \qquad x_1 = 2x_s'$$

$$x_s = \frac{2}{3}x_1, \qquad x_1 = 3x_s'$$

Then

$$\frac{U_1}{U_0} = 1 - \frac{1}{1 + \left(\dfrac{y}{x_s'}\right)^2}$$

$$\frac{U_2}{U_0} = 1 - \frac{3}{4 + \left(\dfrac{y}{x_s'}\right)^2}$$

$$\frac{U_3}{U_0} = 1 - \frac{5}{9 + \left(\dfrac{y}{x_s'}\right)^2}$$

Values of $1 + (y/x_s')^2$, $4 + (y/x_s')^2$ and $9 + (y/x_s')^2$ can be read from Example 3.1 and the following table of U_n/U_0 ($n = 1, 2, 3$) obtained:

y/x_s'	0	0.5	1.0	1.5	2.0	2.5	3.0	3.5	4.0
U_1/U_0	0	0.200	0.500	0.692	0.800	0.862	0.900	0.925	0.941
U_2/U_0	0.250	0.295	0.400	0.520	0.625	0.708	0.769	0.815	0.850
U_3/U_0	0.445	0.460	0.500	0.555	0.615	0.672	0.722	0.765	0.800

These functions are sketched in Fig. 3.20.

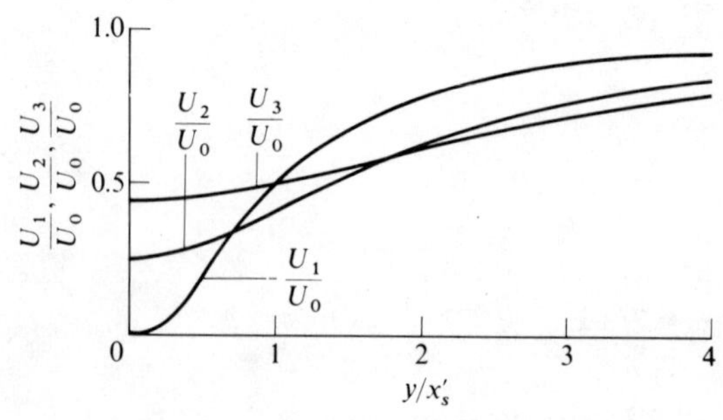

Fig. 3.20. Distribution of discharge velocity downstream of a well near the shoreline of a lake.

(b) From equation 3.52

$$-\frac{q}{\pi x_1 U_0} = 1$$

$$x_1 = 30 \text{ m}$$

$$U_0 = -3.5 \times 10^{-8} \text{ m/s}$$

$$q = \pi \times 30 \times 3.5 \times 10^{-8} \text{ m}^2/\text{s}$$

$$= 3.30 \times 10^{-6} \text{ m}^2/\text{s}$$

$$\text{Total flow rate} = 12 \times 3.30 \times 10^{-6} \text{ m}^3/\text{s}$$

$$= 3.95 \times 10^{-5} \text{ m}^3/\text{s}$$

$$= 143 \text{ 1/hour}$$

Example 3.3

A confined aquifer is supplied with water by a lake with a straight shoreline. The thickness of the aquifer is 17 m, its permeability K is 2.94×10^{-4} m/s and the gradient of piezometric head is 1 in 1420, falling from the lake.

(a) Determine the length of shoreline which contributes water to a fully penetrating well 35 m from the shoreline when water is pumped from the well at the rate of 620 1/hour.

(b) Sketch the distribution along the shoreline of the velocity of the water entering the aquifer.

From equation 3.59

$$\tan\left(1 - \frac{2U_0 y}{q}\right)\frac{\pi}{2} = \frac{y}{x_1}$$

where y is half the desired length

$$\frac{\pi}{2} - \frac{\pi U_0 x_1}{q} \frac{y}{x_1} = \tan^{-1}\left(\frac{y}{x_1}\right)$$

$$x_1 = 35 \text{ m}$$

$$U_0 = -2.94 \times 10^{-4} \times \left(-\frac{1}{1420}\right)$$

$$= 2.07 \times 10^{-7} \text{m/s}$$

$$q = \frac{620 \times 10^{-3}}{17 \times 3600} \text{ m}^2/\text{s}$$

$$= 1.014 \times 10^{-5} \text{m}^2/\text{s}$$

$$\frac{\pi U_0 x_1}{q} = \frac{\pi \times 2.07 \times 10^{-7} \times 35}{1.014 \times 10^{-5}} = 2.24$$

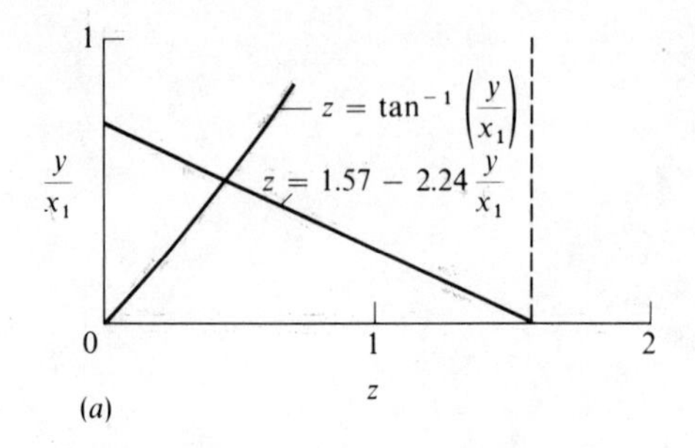

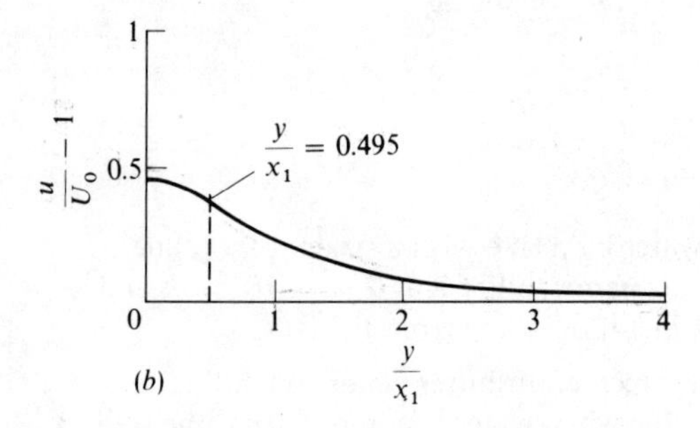

Fig. 3.21. Flow from a lake to a well. (*a*) Solution of $1.57-2.24y/x_1 = \tan^{-1}(y/x_1)$. (*b*) Distribution of discharge velocity entering aquifer from lake.

To find y, solve

$$1.57 - 2.24\frac{y}{x_1} = \tan^{-1}\left(\frac{y}{x_1}\right)$$

From Fig. 3.21

$$\frac{y}{x_1} = 0.495$$

Check: $1.57 - 2.24 \times 0.495 = 0.46$

$\tan^{-1} 0.495 = 0.46$

Hence,

$$2y = 2 \times 0.495 \times 35$$

$$= 34.6 \text{ m}$$

From equation 3.57, with $x = 0$

$$u = U_0 - \frac{q}{2\pi}\left[-\frac{2x_1}{x_1^2 + y^2}\right]$$

$$\frac{u}{U_0} = 1 + \frac{q}{\pi U_0 x_1}\,\frac{1}{1 + \left(\frac{y}{x_1}\right)^2}$$

$$\frac{u}{U_0} = 1 + \frac{0.446}{1 + \left(\dfrac{y}{x_1}\right)^2}$$

$\dfrac{y}{x_1}$	0.	0.5	1.0	1.5	2.0	2.5	3.0	3.5	4.0
$1 + \left(\dfrac{y}{x_1}\right)^2$	1	1.25	2	3.25	5	7.25	10	13.25	17
$\dfrac{u}{U_0}$	1.446	1.358	1.224	1.137	1.089	1.062	1.045	1.034	1.026

The function $\dfrac{u}{U_0} - 1 = f\left(\dfrac{y}{x_1}\right)$ is sketched in Fig. 3.21.

4
Flow nets derived from curvilinear coordinates

Seepage under dams and the associated uplift pressures comprise a very important class of confined flow problems. They are, however, more difficult to analyse than those we have dealt with so far. The analytical difficulties arise because the boundaries are usually described by functions which are not simple and this means that more powerful mathematical tools become necessary. With appropriate transformations, a large number of problems can be solved, although a considerable amount of work is often required. However, the essence of problems of this kind can be discussed without the use of transformation methods.

4.1 Curvilinear coordinates and flow nets

The use of plane polar coordinates as an alternative to cartesian coordinates will be familiar. Using them, we describe the position of a point with the coordinates (r, θ) instead of the cartesian coordinates (x, y). The transformation between the two is made by the equations

$$x = r \cos \theta$$

$$y = r \sin \theta \qquad 4.1$$

To make this concept more general, let there be a system of curvilinear coordinates such that the position of a point is (α, β) and let

$$x = f(\alpha, \beta)$$

$$y = g(\alpha, \beta) \qquad 4.2$$

It follows that α and β are both functions of x and y, although the functions might be difficult to define. That is, we can think of α and β as quantities whose distributions over the $x\,y$ plane are given by equation 4.2.

A network of contours of α and β could be plotted and the network could be a flow net if the two sets of contours were to intersect orthogonally. Then, the contours of α could be equipotentials and the contours of β could be streamlines, or vice versa. For this to be possible, the Cauchy-Riemann conditions must be satisfied:

$$\frac{\partial\alpha}{\partial x}=\frac{\partial\beta}{\partial y}$$

$$\frac{\partial\beta}{\partial x}=-\frac{\partial\alpha}{\partial y}. \qquad 4.3$$

Since α and β may not be available explicitly as functions of x and y, the differentials in equation 4.3 must be found as follows: from equation 4.2

$$\mathrm{d}x=\frac{\partial x}{\partial\alpha}\,\mathrm{d}\alpha+\frac{\partial x}{\partial\beta}\,\mathrm{d}\beta$$

$$\mathrm{d}y=\frac{\partial y}{\partial\alpha}\,\mathrm{d}\alpha+\frac{\partial y}{\partial\beta}\,\mathrm{d}\beta. \qquad 4.4$$

Solving equation 4.4 for $\mathrm{d}\alpha$ and $\mathrm{d}\beta$ yields

$$\mathrm{d}\alpha=\frac{1}{J}\frac{\partial y}{\partial\beta}\,\mathrm{d}x-\frac{1}{J}\frac{\partial x}{\partial\beta}\,\mathrm{d}y \qquad 4.5$$

and

$$\mathrm{d}\beta=-\frac{1}{J}\frac{\partial y}{\partial\alpha}\,\mathrm{d}x+\frac{1}{J}\frac{\partial x}{\partial\alpha}\,\mathrm{d}y$$

where the Jacobian is given by

$$J=\begin{vmatrix}\dfrac{\partial x}{\partial\alpha} & \dfrac{\partial x}{\partial\beta}\\[2mm] \dfrac{\partial y}{\partial\alpha} & \dfrac{\partial y}{\partial\beta}\end{vmatrix} \qquad 4.6$$

But

$$\mathrm{d}\alpha=\frac{\partial\alpha}{\partial x}\,\mathrm{d}x+\frac{\partial\alpha}{\partial y}\,\mathrm{d}y \qquad 4.7$$

and

$$\mathrm{d}\beta=\frac{\partial\beta}{\partial x}\,\mathrm{d}x+\frac{\partial\beta}{\partial y}\,\mathrm{d}y$$

Comparison of equations 4.5 and 4.7 yields

$$\frac{\partial\alpha}{\partial x}=\frac{1}{J}\frac{\partial y}{\partial\beta} \qquad \frac{\partial\alpha}{\partial y}=-\frac{1}{J}\frac{\partial x}{\partial\beta}$$

$$\frac{\partial\beta}{\partial x}=-\frac{1}{J}\frac{\partial y}{\partial\alpha} \qquad \frac{\partial\beta}{\partial y}=\frac{1}{J}\frac{\partial x}{\partial\alpha} \qquad 4.8$$

Hence, the Cauchy-Riemann conditions will be satisfied if

$$\frac{\partial x}{\partial \alpha} = \frac{\partial y}{\partial \beta}$$

$$\frac{\partial x}{\partial \beta} = -\frac{\partial y}{\partial \alpha} \qquad 4.9$$

Now consider, as an example

$$x = a \cosh \alpha \cos \beta$$

$$y = a \sinh \alpha \sin \beta. \qquad 4.10$$

Differentiation will show that equations 4.9 are satisfied, so that α and β could represent a potential and a stream function. Let α be a dimensionless potential defined by

$$\alpha = \frac{\alpha_0 \phi}{K \Delta H} \qquad 4.11$$

and let β be a dimensionless stream function defined by

$$\beta = \frac{\beta_0 \psi}{\Delta q} \qquad 4.12$$

Here, K is the permeability of the soil, ΔH is a known difference in potential between two equipotentials and Δq is a known discharge between two streamlines. The range of values of α is from 0 to α_0 corresponding to the range of potential from 0 to $K\Delta H$ and β varies similarly from 0 to β_0 for a range of stream function from 0 to Δq. Fig. 4.1 shows diagrammatically how the ranges of α and ϕ are to be divided, so that there is a corresponding contour of ϕ for every contour of α.

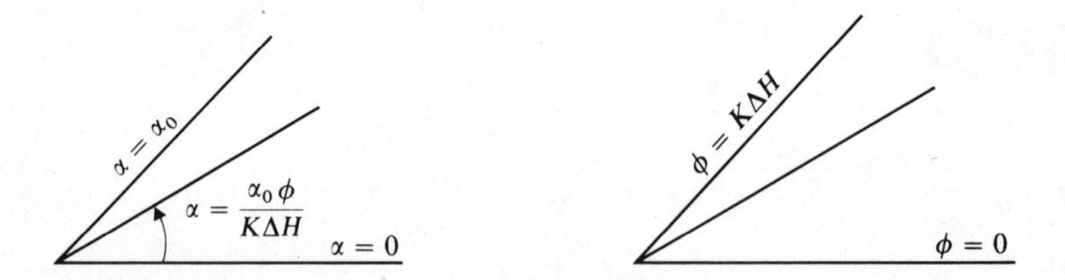

Fig. 4.1. Illustration of the application of equation 4.11. A similar sketch would apply to equation 4.12.

The contours are defined by equation 4.10, each value of ϕ giving a particular curve in the x y plane and similarly for each value of ψ (equations 4.10, 4.11, 4.12). The equipotentials are found by eliminating β from equation 4.10, and the streamlines by eliminating α as follows:

$$\frac{x^2}{(a \cosh \alpha)^2} = \cos^2 \beta$$

$$\frac{y^2}{(a \sinh \alpha)^2} = \sin^2 \beta$$

adding both sides yields

$$\frac{x^2}{(a \cosh \alpha)^2} + \frac{y^2}{(a \sinh \alpha)^2} = 1 \qquad 4.13$$

Likewise

$$\frac{x^2}{(a \cos \beta)^2} = \cosh^2 \alpha$$

$$\frac{y^2}{(a \sin \beta)^2} = \sinh^2 \alpha$$

and

$$\frac{x^2}{(a \cos \beta)^2} - \frac{y^2}{(a \sin \beta)^2} = 1 \qquad 4.14$$

Equation 4.13 is a family of confocal ellipses and equation 4.14 a family of hyperbolas, as shown in Fig. 4.2.

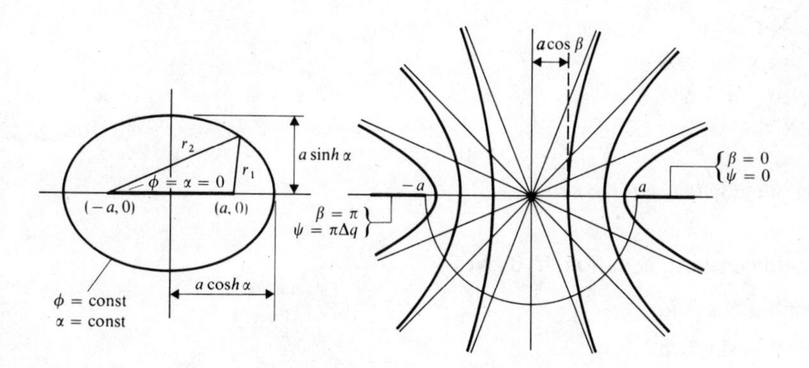

Fig. 4.2. Illustration of equation 4.13 and 4.14.

The velocity components at any point in this flow net may be found from $u = -\partial\phi/\partial x$, $v = -\partial\phi/\partial y$ as follows:

$$u = -\frac{\partial \phi}{\partial x} = -\frac{K\Delta H}{\alpha_0}\frac{\partial \alpha}{\partial x} = -\frac{K\Delta H}{J\alpha_0}\frac{\partial y}{\partial \beta}$$

and

$$v = -\frac{\partial \phi}{\partial y} = -\frac{K\Delta H}{\alpha_0}\frac{\partial \alpha}{\partial y} = \frac{K\Delta H}{J\alpha_0}\frac{\partial x}{\partial \beta} \qquad 4.15$$

The Jacobian is evaluated by means of equation 4.6 and we obtain

$$J = a^2 (\sinh^2 \alpha \cos^2 \beta + \cosh^2 \alpha \sin^2 \beta)$$

$$= \frac{a^2}{2} (\cosh 2\alpha - \cos 2\beta). \quad 4.16$$

Substituting this in equation 4.15 gives

$$u = -\frac{2 K\Delta H \sinh \alpha \cos \beta}{\alpha_0 a (\cosh 2\alpha - \cos 2\beta)}$$

$$v = -\frac{2 K\Delta H \cosh \alpha \sin \beta}{\alpha_0 a (\cosh 2\alpha - \cos 2\beta)}. \quad 4.17$$

In an alternative approach to the calculation of the velocity at any point, we take the orthogonality of the confocal ellipses and hyperbolas as proved. Then, with $\alpha = \alpha_0 \phi / K\Delta H$ the velocity vector **q** is given by

$$\mathbf{q} = -\frac{K\Delta H}{\alpha_0} \frac{\partial \alpha}{\partial n} \quad 4.18$$

where Δn is an element of length of a streamline (along which β is constant), Fig. 4.3.

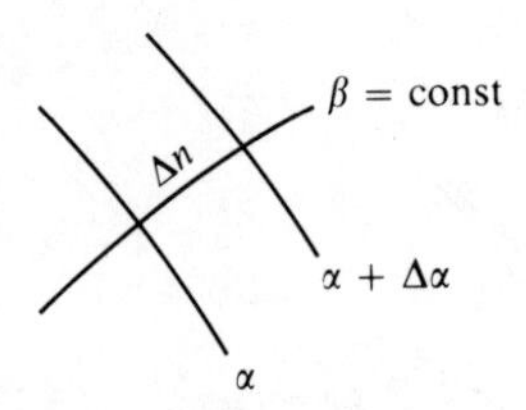

Fig. 4.3. Definition sketch for α, β and Δn.

With β held constant, equation 4.10 gives

$$dx = a \sinh \alpha \cos \beta \, d\alpha$$

$$dy = a \cosh \alpha \sin \beta \, d\alpha$$

whence

$$\Delta n = a \sqrt{\cosh^2 \alpha - \cos^2 \beta} \, \Delta \alpha$$

so that

$$\frac{\partial \alpha}{\partial n} = \frac{1}{a \sqrt{\cosh^2 \alpha - \cos^2 \beta}}$$

Substituting in equation 4.18 gives

$$q = -\frac{K\Delta H}{\alpha_0 a \sqrt{\cosh^2 \alpha - \cos^2 \beta}} \quad 4.19$$

This result can be interpreted geometrically in the following way. The point where we have found the velocity is at the intersection of a certain ellipse (the equipotential, α = constant) and a hyperbola (the streamline, β = constant). If r_1 and r_2 (Fig. 4.2) are the distances of this point from the foci, then the major axis of the ellipse = $2a \cosh \alpha = r_1 + r_2$, the major axis of the hyperbola = $2a \cos \beta = r_1 - r_2$, and

$$q = -\frac{K\Delta H}{\alpha_0 \sqrt{\left(\frac{r_1 + r_2}{2}\right)^2 - \left(\frac{r_1 - r_2}{2}\right)^2}} \qquad 4.20$$

$$= -K \frac{\Delta H}{\alpha_0 \sqrt{r_1 r_2}}$$

Notice that near the ends of the $\alpha = 0$ ellipse, $r_1 \to 0$ so that theoretically infinite velocities occur.

All of the analysis above is directly applicable to the alternative flow net in which the hyperbolas are the equipotentials and the ellipses are the streamlines. All that is required is an alteration to the definitions of α and β

$$\alpha = \frac{\alpha_0 \psi}{\Delta q}$$

$$\beta = \frac{\beta_0 \phi}{K\Delta H} \qquad 4.21$$

In the following section flow nets as discussed here are applied to problems of seepage under or around structures and from canals.

4.2 Seepage under an impervious dam

The analysis above can be applied to this problem by fitting the physical boundaries to the flow net. In this case, we require the ellipses to be the streamlines and the hyperbolas to be the equipotentials

$$\alpha = \frac{\alpha_0 \psi}{\Delta q} \quad \text{and} \quad \beta = \frac{\beta_0 \phi}{K\Delta H} \qquad 4.22$$

In Fig. 4.4 the base of the dam **AB** is a plane resting on top of an infinite porous medium. The base is obviously a streamline. To the left of the base, the top surface of the porous medium is the equipotential corresponding to the head H_1 and to the right, it is another equipotential which corresponds to the head H_2. Referring to Fig. 4.4 it will be seen that the x axis between the foci ($-a < x < a$), where $\alpha = 0$ is one of the family of ellipses (streamlines) while for $x < -a$, the x axis is a hyperbola given by $\beta = \pi$ and for $x > a$ it is a hyperbola given by $\beta = 0$. Equation 4.22 lead us to the following conclusions about the top surface of the porous medium: upstream of the dam, equipotential $\phi = K\Delta H$: under the dam, streamline, $\psi = 0$; downstream of the dam, equipotential $\phi = 0$. We also see that the range of values of β is between 0 and π so that $\beta_0 = \pi$. Hence

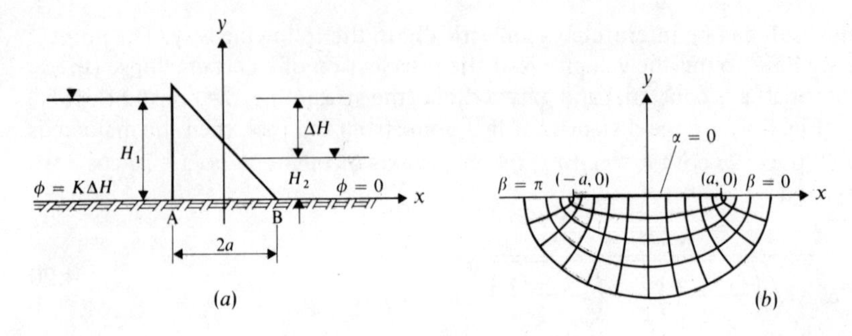

Fig. 4.4. Definition sketch for seepage under a flat-based dam.

from the second of equation 4.22

$$\beta = \frac{\pi\phi}{K\Delta H} \tag{4.23}$$

The question of interest now is the distribution of pressure under the base of the dam. We have

$$x = a \cosh \alpha \cos \beta$$

and, along AB, $-a < x < a$ and $\alpha = 0$

$$x = a \cos \beta$$

$$\beta = \cos^{-1} \frac{x}{a} \tag{4.24}$$

Using equation 4.23

$$\phi = K\Delta H \frac{\cos^{-1} \frac{x}{a}}{\pi} \tag{4.25}$$

Equation 3.2 may be introduced now yielding

$$Kh + C = \frac{1}{\pi} K\Delta H \cos^{-1} \frac{x}{a}$$

and substituting $h = H_2$ when $x = a$ gives

$$C = -KH_2$$

The pressure distribution (Fig. 4.5) is therefore given by

$$h = H_2 + \frac{1}{\pi} \Delta H \cos^{-1} \frac{x}{a} \tag{4.26}$$

From here, the resultant upward force and the overturning moment may be

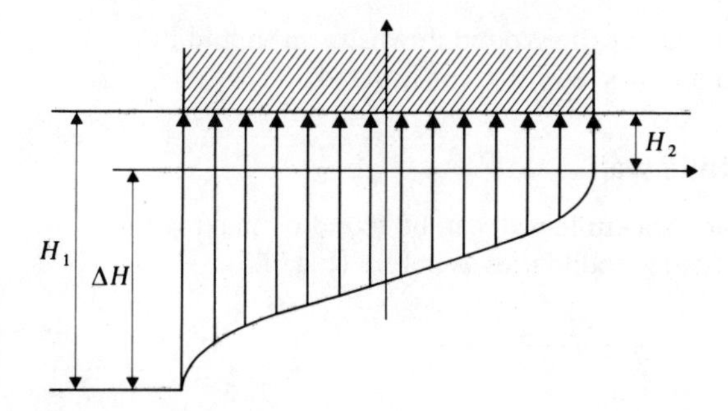

Fig. 4.5. Distribution of uplift under a dam.

calculated by straightforward integrations, as follows:

$$F = \int_{-a}^{a} \gamma h \, dx$$

$$= \gamma \int_{-a}^{a} \left(H_2 + \frac{1}{\pi} \Delta H \cos^{-1} \frac{x}{a} \right) dx$$

$$= \gamma \left(H_2 + \frac{\Delta H}{2} \right) 2a \qquad 4.27$$

and

$$M = \int_{-a}^{a} (a - x) \gamma h \, dx$$

$$= \gamma \int_{-a}^{a} (a - x) \left(H_2 + \frac{1}{\pi} \Delta H \cos^{-1} \frac{x}{a} \right) dx$$

$$= \gamma \frac{a^2}{4} (5H_2 + 3H_1)$$

What is of greater importance is that the pressure gradient is infinitely large at both $x = -a$ and $x = +a$. On the upstream side, this is not significant, since the pressure force on the grains in the permeable base is directed downwards. At the downstream end, however, the pressure force acts upwards and can lead to failure of the ground. For cohesionless ground the granular medium becomes 'quicksand' when the pressure gradient is steeper than the floatation gradient $i_f = (1 - n)(S_s - 1)$. Here n is the porosity and S_s the specific gravity of the grains. There is no such simple relationship for cohesive soils. In practice, infinitely large gradients and forces do not occur. The analysis based on Darcy's law breaks down and no longer describes the physical situation accurately when the Reynolds number gets too large. Using the correct seepage equation would yield finite gradients. Nevertheless, the analysis does show the downstream toe is a danger zone. In order to make the structure safe, the base of the dam should be lowered below the top of the porous medium, a cut-off wall should be constructed at the downstream side or the ground downstream should be weighted with overburden and an inverted filter constructed.

4.3 Seepage under a sheetpile wall

The flow net for seepage under a sheetpile wall can be found by making the x axis coincide with wall and fitting boundaries as before (Fig. 4.6).

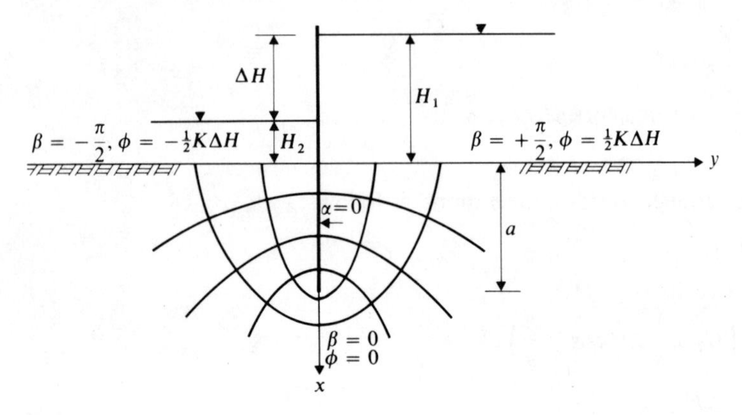

Fig. 4.6. Flow past a sheetpile cut-off wall.

Along the y axis, $x = 0$, so that

$$a \cosh \alpha \cos \beta = 0$$

whence

$$\beta = \pm \frac{\pi}{2}$$

Along the x axis, $y = 0$ and

$$a \sinh \alpha \sin \beta = 0$$

Thus, either $\beta = 0$ or $\alpha = 0$. The first of these gives $\cosh \alpha = x/a$ (equation 4.10) and is therefore possible only for $x > a$, i.e. below the tip of the sheetpile wall. The second gives $\cos \beta = x/a$ (equation 4.10) and is possibly only for $x < a$. It will be seen (Fig. 4.6) that the range of head ΔH corresponds to a range of β equal to π. Hence, the transformation from β to ϕ is equation 4.22

$$\beta = \frac{\pi \phi}{K \Delta H} \qquad 4.28$$

Consequently, on the y axis

$$\phi = \pm \tfrac{1}{2} K \Delta H \qquad 4.29$$

and on the x axis below the tip of the wall

$$\phi = 0$$

Along the sheetpile wall, $0 < x < a$ and $\alpha = 0$. Hence, from equation 4.10

$$\cos \beta = \frac{x}{a} \qquad 4.30$$

The velocity of flow along the wall is

$$\begin{aligned} u_{y=0} &= -\left.\frac{\partial \phi}{\partial x}\right|_{\alpha=0} \\ &= -\frac{K\Delta H}{\pi}\left.\frac{\partial \beta}{\partial x}\right|_{\alpha=0} \\ &= \frac{K\Delta H}{\pi J}\left.\frac{\partial y}{\partial \alpha}\right|_{\alpha=0} \\ &= \frac{K\Delta H}{\pi a}\frac{1}{\sin \beta} \\ &= \pm\frac{K\Delta H}{\pi a}\frac{1}{\sqrt{1-\left(\frac{x}{a}\right)^2}} \end{aligned} \qquad 4.31$$

The positive value is for $0 < \beta < +\pi/2$, as on the right side of the wall in Fig. 4.6 and the negative value for $-\pi/2 < \beta < 0$, as on the left side. The velocities at entry and exit are found by calculating u for $x = 0$ and $\beta = \pm\pi/2$. Equation 4.10 gives

$$\sinh \alpha = \pm\frac{y}{a}$$

Also,

$$\begin{aligned} u_{x=0} &= -\left.\frac{\partial \phi}{\partial x}\right|_{\beta=\pm\frac{\pi}{2}} \\ &= \pm\frac{K\Delta H}{\pi a}\frac{1}{\sinh \alpha} \\ &= \pm\frac{K\Delta H}{\pi a}\frac{1}{\sqrt{1+\left(\frac{y}{a}\right)^2}} \end{aligned} \qquad 4.32$$

Here, the positive value is at entry ($\beta = +\pi/2$) and the negative value is at exit ($\beta = -\pi/2$). These velocities are shown in Fig. 4.7.

The steepest pressure gradient at exit is an important quantity which may be found from the maximum velocity at exit, as follows:

$$(u_{x=0})_{max} = -\frac{K\Delta H}{\pi a} = -Ki_{max}$$

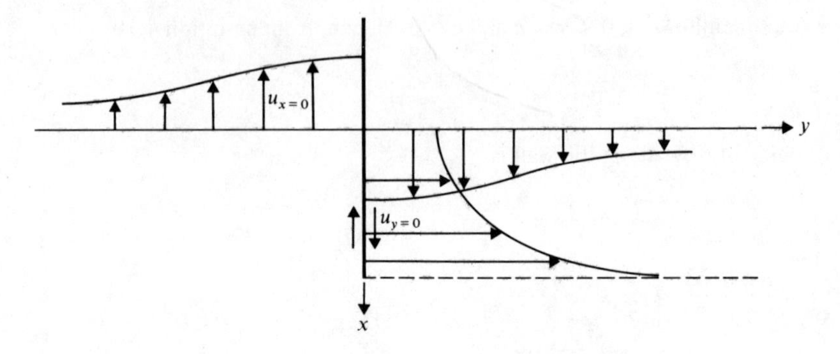

Fig. 4.7. Velocity distributions in flow past a sheetpile cut-off wall.

Hence,

$$i_{max} = \frac{\Delta H}{\pi a} \qquad 4.33$$

and this should be less than the floatation gradient i_f. Equation 4.33 is used to calculate the penetration a which is required for a given floatation gradient, factor of safety and head differential.

For the pressure distribution on the sheet piling we have

$$\alpha = 0, \text{ whence } \cos\beta = \frac{x}{a}$$

$$\beta = \frac{\pi\phi}{K\Delta H}, \text{ whence } \phi = \frac{K\Delta H}{\pi}\beta$$

and

$$\phi = Kh + C \text{ whence}$$

$$Kh + C = \frac{K\Delta H}{\pi}\cos^{-1}\frac{x}{a}$$

When $x = 0$ and $\beta = +\pi/2$ (i.e. at entry) $h = H_1$

$$KH_1 + C = ½K\Delta H$$

Hence

$$H_1 - h = ½\Delta H - \frac{H}{\pi}\cos^{-1}\frac{x}{a}$$

$$h = H_1 - \Delta H\left(\frac{\pi}{2} - \cos^{-1}\frac{x}{a}\right) \qquad 4.34$$

Along the right side of the wall, $\cos^{-1} x/a$ (= β) is between 0 and $+\pi/2$ and on the left side, it is between 0 and $-\pi/2$. This pressure distribution is shown in Fig. 4.8.

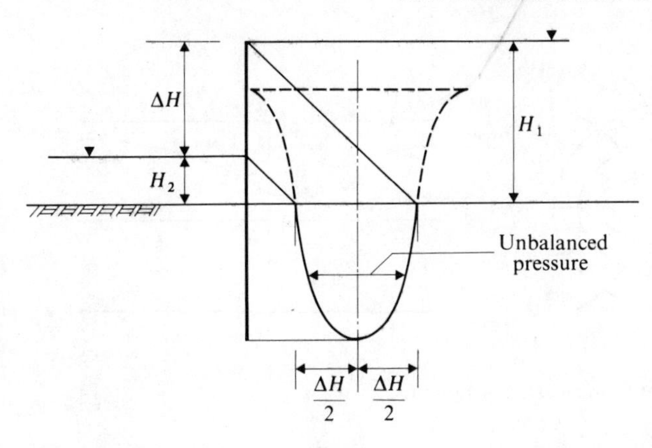

Fig. 4.8. Pressure distribution for a sheetpile cut-off wall.

It must be emphasized that the pressure distribution referred to here is the hydrodynamic pressure and does not include the soil pressure. For a complete analysis of the total load on a sheet-pile wall, reference could be made to Rowe (1952, 1954, 1955, 1957, 1963).

4.4 Seepage from or to a canal or a slot

If the confocal ellipses are assumed to be equipotential lines and the hyperbolas the streamlines, the corresponding physical problem could be seepage through the pervious bed of a canal or to or from a slot (Fig. 4.9, Fig. 4.10).

Along the slot, or the bottom of the canal, $\alpha = 0$, $y = 0$, $-a < x < a$, and $x = a \cos \beta$. For the rest of the x axis, $\beta = 0$ for $x > a$ and π for $x < -a$ for flow from the canal. If the flow is to the canal, these values of β are reversed. Thus, from one bounding streamline to the other, β ranges from 0 to π. Hence $\beta_0 = \pi$ and if Δq is the discharge through the bed of the canal

$$\beta = \frac{\pi\psi}{\Delta q} \tag{4.35}$$

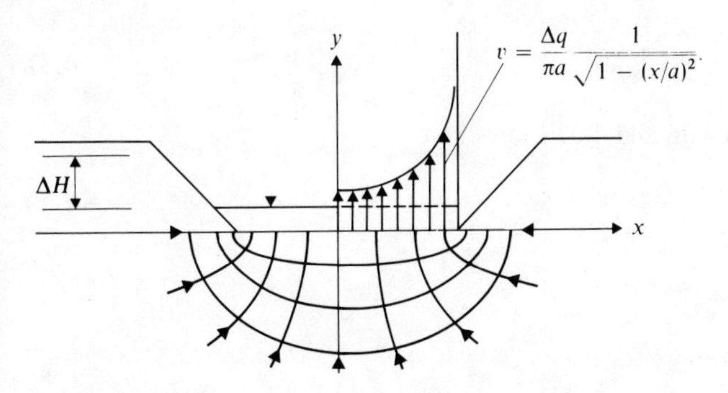

Fig. 4.9. Flow through a slot from an artesian aquifer.

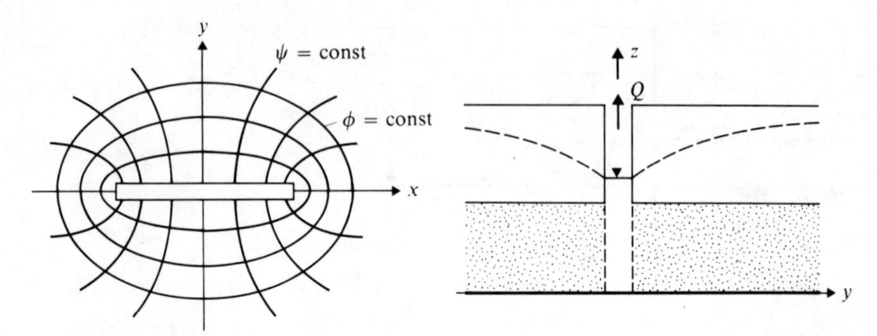

Fig. 4.10. Flow into a slot from an artesian aquifer.

The velocity of flow across the x axis is

$$v_{y=0} = \left.\frac{\partial\psi}{\partial x}\right|_{y=0} = \left.\frac{\Delta q}{\pi}\frac{\partial\beta}{\partial x}\right|_{y=0}$$

using equation 4.35. Substituting $\frac{\partial\beta}{\partial x} = -\frac{1}{J}\frac{\partial y}{\partial\alpha}$, $y = a \sinh\alpha \sin\beta$ and $\alpha = 0$ yields

$$v_{y=0} = -\frac{\Delta q}{\pi a}\frac{1}{\sin\beta} = -\frac{\Delta q}{\pi a}\frac{1}{\sqrt{1-\left(\frac{x}{a}\right)^2}} \qquad 4.36$$

The distribution of potential in the aquifer may be found by calculating the velocity across the x axis from

$$v_{y=0} = -\left.\frac{\partial\phi}{\partial y}\right|_{y=0} = -\frac{K\Delta H}{\alpha_0}\frac{\partial\alpha}{\partial y}$$

using $\alpha = \alpha_0\phi/K\Delta H$. Substituting $\frac{\partial\alpha}{\partial y} = -\frac{1}{J}\frac{\partial x}{\partial\beta}$, $x = a\cosh\alpha\cos\beta$ and $\alpha = 0$ yields

$$v_{y=0} = -\frac{K\Delta H}{a\alpha_0}\frac{1}{\sin\beta} \qquad 4.37$$

Comparing equations 4.36 and 4.37 shows that

$$\alpha_0 = \frac{\pi K\Delta H}{\Delta q} \qquad 4.38$$

whence

$$\phi = \frac{\Delta q}{\pi}\alpha \qquad 4.39$$

If it can be assumed that the drawdown ΔH occurs over a large distance r_e, a circle of radius r_e is a close approximation for the equipotential $\alpha = \alpha_0$. Large values of α make $\cosh \alpha = \sin \alpha = \frac{1}{2}e^{\alpha}$, and it can be shown that $r_e = ae^{\alpha_0}/2$, whence

$$\alpha_0 = \ln \frac{2r_e}{a} \tag{4.40}$$

Comparing equation 4.38 and 4.40 we get

$$q = \frac{\pi K \Delta H}{\ln \frac{2r_e}{a}} \tag{4.41}$$

It is of interest that this is half the discharge to a well of radius $a/2$ if the drawdown is ΔH.

4.5 Remarks

In the cases of the dam and the canal the permeable medium has been assumed homogeneous and semi-infinite. In the case of the slot, an infinite medium has been assumed. For the case of a finite depth of porous medium under the dam as well as for the effect of cut-off walls, the reader is referred to the work of Pavlovsky which has been reproduced in English by Leliavsky (1955). The following references are also recommended for additional reading: Cedergren (1967), Muskat (1937), Khosla *et al.* (1936), Polubarinova-Kochina (1962). With respect to cut-off walls, it should be noted that they can reduce the total uplift by increasing the length of the seepage path if located at the upstream end of the base of the dam. However, this should be taken into account only if the permeability of the cut-off is definitely known. It emerged from the investigations for the design of Roxburgh dam in New Zealand that the permeability of a cut-off, with depth half the base width of the dam, must be less than 5% of the permeability of the foundation to get a significant reduction in uplift (Fig. 4.11, Raudkivi, (1954)).

The flow net derived from such an elementary model as the flat based dam does not, of course, give a solution for a complex practical problem. However, it does serve to indicate the nature of the pattern. This knowledge is useful when

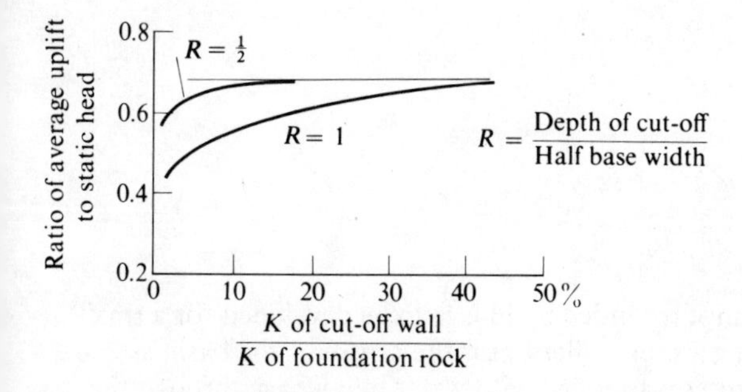

Fig. 4.11. Effect of relative permeability of cut-off on uplift.

a flow net has been sketched or found by experiment. Once a flow net has been obtained, the rate of seepage, the pressures, pressure gradients, etc. can be determined from the net.

The flow rate per unit width of structure is

$$Q = n_s \Delta q = n_s K \Delta h$$

where n_s is the number of streamtubes. Writing for the number of equipotential drops (i.e. the number of squares in the direction of flow) $n_p = H/\Delta h$ yields on substitution

$$Q = \frac{n_s}{n_p} KH$$

At any point in the flow net, the value of h can be determined by simply counting down from the head at entry H. The pressure follows from $h = z + p/\gamma$. The spacing of the equipotentials is a measure of the pressure gradient – the potential drop is constant between any pair of adjacent equipotentials, so that the closer they are, the steeper the gradient.

Example 4.1

A sheetpile cofferdam is required to enclose a construction site in water 11 m deep. The bed material is sand of specific gravity 2.5 and its porosity is 35%.

Estimate the depth to which the sheetpiles must be driven if a factor of safety of 1.6 is required with respect to the floatation gradient.

$$\text{Floatation gradient} = (S_s - 1)(1 - n)$$

$$= (2.5 - 1)(1 - 0.35)$$

$$= 0.975$$

$$i_{max} = \frac{0.975}{1.6} = 0.61$$

From equation 4.33

$$a = \frac{\Delta H}{\pi i_{max}}$$

$$= \frac{11}{\pi \times 0.61}$$

$$= 5.75 \text{ m}$$

Example 4.2

A confined aquifer in a basin surrounded by hills is to be developed for a small water supply by constructing a short gallery near the centre of the basin as shown in Fig. 4.12. The aquifer is supplied with water by seepage through the upper confining layer where it is fractured remote from the site of the gallery

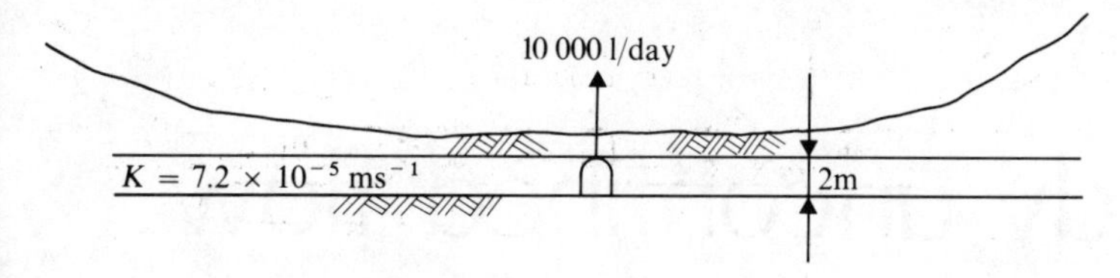

Fig. 4.12. Definition sketch for flow to a gallery.

and it is estimated that the piezometric head at a radius of 1.1 km from the centre of the basin will be unchanged by the development.

The thickness of the aquifer is 2 m and its permeability K is 7.2 x 10^{-5} m/s. Estimate the length of gallery required to supply 10 000 l per day if the drawdown is limited to 1.5 m.

From equation 4.41

$$\ln \frac{2r_e}{a} = \frac{\pi K \Delta H}{q}$$

$$q = \frac{10\,000 \times 10^{-3}}{86\,400 \times 2} = 5.787 \times 10^{-5} \text{m}^2/\text{s}$$

$$K = 7.2 \times 10^{-5} \text{m/s}$$

$$\Delta H = 1.5 \text{ m}$$

$$\ln \frac{2r_e}{a} = \frac{\pi \times 7.2 \times 10^{-5} \times 1.5}{5.787 \times 10^{-5}}$$

$$= 5.863$$

$$\frac{2r_e}{a} = 351.766$$

$$2r_e = 2.2 \times 10^3 \text{ m}$$

$$a = 6.254 \text{ m}$$

$$2a = 12.51 \text{ m}$$

Estimated length of gallery required – say 15 m.

Note that the equipotentials are confocal ellipses with foci 15 m apart. However, at a distance of 1.1 km from the gallery, the equipotential ellipse is very close to a circle.

5
Steady unconfined flow

The class of problems to be introduced here is characterized by the feature that a part of the boundary defining the flow is a free surface of initially unknown shape. The free surface is a streamline, or stream surface, along which the pressure is uniform and the shape is such that the equations of motion and continuity are satisfied everywhere. This property makes such flows inherently more difficult to analyse than confined flows, where the geometry of all boundaries is known in advance.

5.1 Surface of seepage

Additional complications arise in gravity flow systems in the form of surfaces of seepage which are neither streamlines or equipotentials. They are simply parts of the boundary of the porous medium, e.g. BC in Fig. 5.1, through which the fluid leaves it. Like the shape of the free surface, the length of the surface of seepage is unknown initially.

It is not difficult to see why such a surface must occur. Assume that EF is the downstream slope of an earth dam or stopbank, with downstream water level at C. The part of the boundary CF below the surface is a constant potential line and the streamline DC must meet it at right angles. The free surface must have a downward slope, as it approaches the outflow surface, and the velocity along this streamline must be finite and, by Darcy's law, directly proportional to the slope of the surface. Assume now that the free surface is defined by A′C. For the velocity along it to be different from zero, the slope at C must be downwards, so that the streamlines DC and A′C converge to the point C. This means that all the flow originating above the streamline DC must discharge through zero area at infinite velocity. Since this is impossible, the free surface will rise until a physically possible velocity, through BC, is attained. The free surface AB should approach EC nearly tangentially because the velocity must be continuous as the water passes from the porous medium into the air at B.

Point C should be noted as a potential danger zone, where the material could become quicksand and be washed away, leading to failure through progressive erosion. Such danger zones can occur at other places, for example as shown in Fig. 5.2.

The shape of the streamlines yields useful information about the variation of

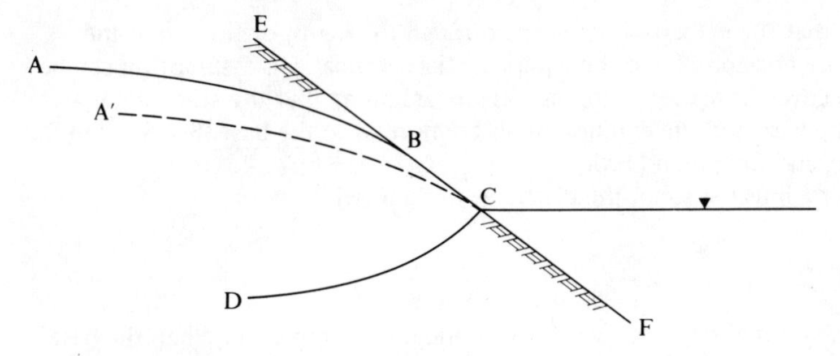

Fig. 5.1. Illustration of surface of seepage.

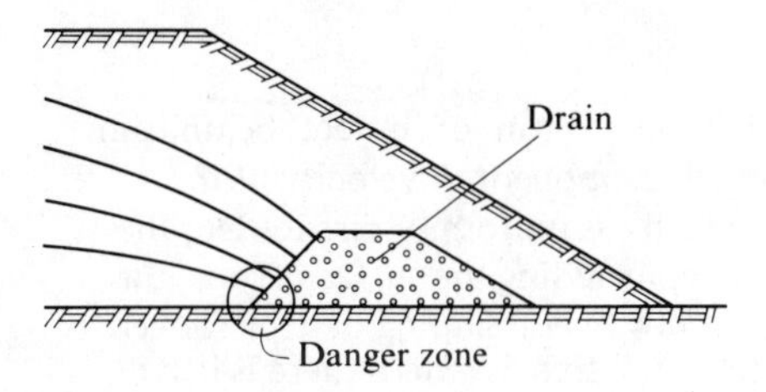

Fig. 5.2. Concentration of streamlines.

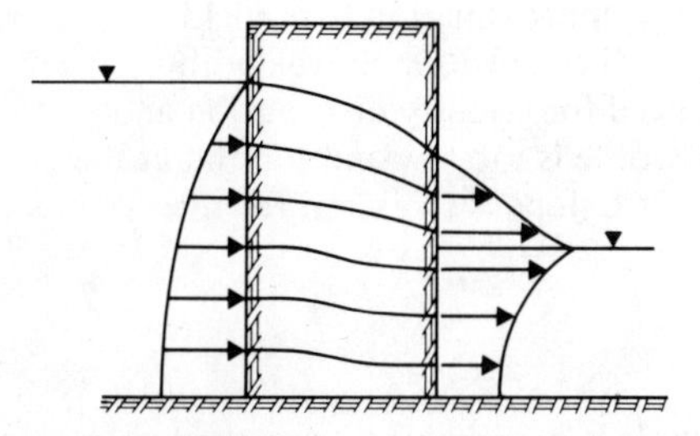

Fig. 5.3. Diagrammatic relationship between streamline curvature and velocity distribution.

velocity with the vertical coordinate, z. In brief, if the streamlines are concave upwards, the velocity increases with z and conversely. (See Fig. 5.3).

We now turn to an analytical study of gravity flow. In considering the continuity requirement, it was shown (equation 2.20) that, under certain conditions, a gravity flow also satisfies the Laplace equation. We now consider this in more detail.

5.2 The Dupuit-Forchheimer approximation

One of the first approximations used for solving unconfined flow problems, and one that is still used extensively, is that due to Dupuit. In 1863, Dupuit formulated the approximation that, for a horizontal base and small inclinations of the free surface, the velocity is proportional to the slope of the water surface.

He assumed that the velocity is constant through the depth of flow, that the streamlines are horizontal and the equipotentials vertical. His assumption can be extended to cover flow over a sloping base by assuming that the streamlines are parallel to the base and the equipotentials are normal to the base. See Section 2.2 and Wooding and Chapman (1966).

Hence, for a linear system, the discharge per unit width is

$$q = -Kz \frac{dh}{dx}$$

where z is the normal depth of water above the impervious base. When the base is horizontal, $dh/dx = dz/dx$ and

$$q = -Kz \frac{dz}{dx} \qquad 5.1$$

It must be realized that the velocity distribution cannot, in fact, be uniform through the depth. For example, the normal component of velocity at the impervious base must be zero and if the velocity is uniformly distributed, the normal component must be zero everywhere, including the surface. Thus, the assumption is strictly true only for parallel flow over a sloping base or for zero flow if the base is horizontal. However, for small slopes, where there is little convergence or divergence of streamtubes, the head loss in the normal direction and the normal component of velocity are negligible in comparison with the head loss and velocity parallel to the base. The approximation then yields reasonable results, particularly for discharge, which correspond well with measurements. The errors are more pronounced for velocity distribution and free surface profile. The calculated surface profile is too low and too flat at the outlet end of the system, where the true surface slope is too steep for the approximation to be valid.

5.3 Linear gravity flow system

The surface profile known as Dupuit's parabola is derived by solving the Laplace equation for flow over a horizontal impervious base. With K constant and no flow in the y direction, the Laplace equation is reduced to

$$\frac{d^2(z^2)}{dx^2} = 0$$

from which

$$z^2 = C_1 x + C_2$$

Assuming boundary conditions, Fig. 5.4, $z = z_1$ when $x = 0$ and $z = z_2$ when $x = L$ yields $C_2 = z_1^2$ and $C_1 = -(z_1^2 - z_2^2)/L$ whence

$$z^2 = z_1^2 - \frac{z_1^2 - z_2^2}{L} x$$

$$\frac{x}{L} = \frac{z_1^2 - z^2}{z_1^2 - z_2^2} \qquad 5.2$$

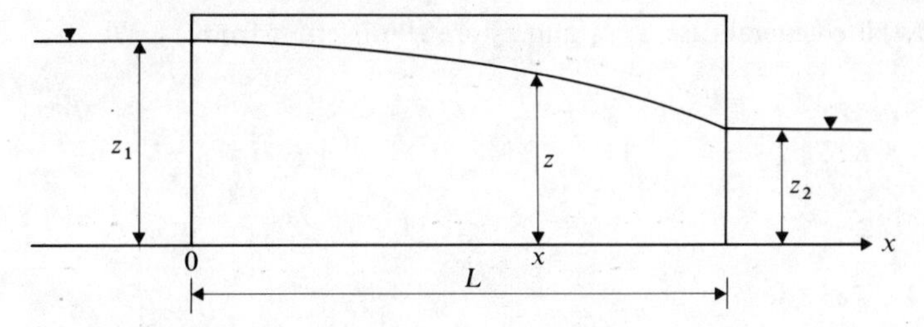

Fig. 5.4. Definition sketch of a linear gravity flow system.

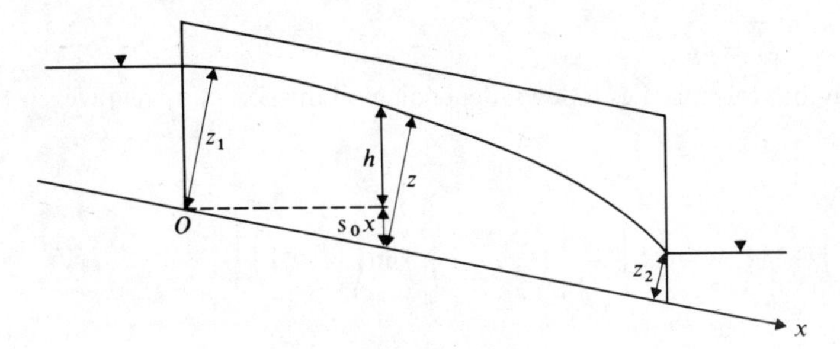

Fig. 5.5. Linear gravity flow system on a sloping impervious layer.

The discharge is given by

$$q = uz = -Kz\frac{dz}{dx} = \frac{K(z_1^2 - z_2^2)}{2L} \tag{5.3}$$

If the impermeable base is not horizontal, but has a slope S_0 defined as positive when the base falls in the direction of flow, then from Fig. 5.5, provided S_0 is small enough, $z = h + S_0x$ and

$$\frac{dh}{dx} = \frac{dz}{dx} - S_0$$

From Darcy's law

$$u = -K\frac{dh}{dx} = -K\frac{dz}{dx} + KS_0$$

and

$$q = -Kz\frac{dz}{dx} + KzS_0$$

If the depth for uniform flow is z_0

$$q = KS_0z_0 \tag{5.4}$$

Let us be quite clear about the interpretation of this depth z_0 – it is the depth of flow required for uniform flow to take place at discharge q and slope S_0. With non-uniform flow, the general case, z, z_1 and z_2 are all different from z_0. We then have

$$KS_0z_0 = -Kz\frac{dz}{dx} + KzS_0$$

whence

$$S_0x = z + z_0 \ln(z - z_0) + C$$

Substituting $x = 0$ and $z = z_1$ to evaluate C yields

$$S_0x = z - z_1 + z_0 \ln\frac{z - z_0}{z_1 - z_0} \qquad 5.5$$

Equation 5.5 may be rearranged as follows, depending on the size of z_1 relative to z_0:

for $z_1 > z_0$

$$\frac{S_0x}{z_0} = \left[\left(\frac{z}{z_0} - 1\right) + \ln\left(\frac{z}{z_0} - 1\right)\right] - \left[\left(\frac{z_1}{z_0} - 1\right) + \ln\left(\frac{z_1}{z_0} - 1\right)\right] \qquad 5.6$$

for $z_1 < z_0$

$$\frac{S_0x}{z_0} = \left[\left(1 - \frac{z_1}{z_0}\right) - \ln\left(1 - \frac{z_1}{z_0}\right)\right] - \left[\left(1 - \frac{z}{z_0}\right) - \ln\left(1 - \frac{z}{z_0}\right)\right] \qquad 5.7$$

Thus once the functions

$$\left[\left(\frac{z}{z_0} - 1\right) + \ln\left(\frac{z}{z_0} - 1\right)\right] \text{ and } \left[\left(1 - \frac{z}{z_0}\right) - \ln\left(1 - \frac{z}{z_0}\right)\right]$$

have been evaluated, any surface profile can be calculated by reading values of these functions from a table. It will also be seen that, if $z_1 > z_0, z > z_1$ – i.e. the flow is expanding – and that if $z_1 < z_0$, the flow is converging. (Compare these observations with the M_1 and M_2 profiles in open channel flow).

In the case of converging flow, if $z_1 \ll z_0$, $\ln(1 - z/z_0)$ can be expanded as follows:

$$\ln\left(1 - \frac{z}{z_0}\right) = -\left(\frac{z}{z_0}\right) - ½\left(\frac{z}{z_0}\right)^2$$

If this is substituted in equation 5.6 and the result simplified

$$\frac{S_0x}{z_0} = ½\left(\frac{z_1}{z_0}\right)^2 - ½\left(\frac{z}{z_0}\right)^2 \qquad 5.8$$

As an indication of the accuracy of this approximation, profiles calculated using equations 5.7 and 5.8 for $z_1/z_0 = 0.2$ are tabulated hereunder:

S_0x/z_0	0	0.002	0.004	0.006	0.008	0.010
z/z_0 (eqn. 5.7)	0.200	0.191	0.182	0.173	0.163	0.153
z/z_0 (eqn. 5.8)	0.200	0.190	0.179	0.167	0.155	0.141

These conclusions are of considerable interest when the potential flow analogy is applied to the problem. We have shown that a potential $\phi = \frac{1}{2}Kz^2 + C$ exists provided $dz/dx \gg \tan\alpha$ or $\alpha \to 0$. The latter condition implies that z_0 will be large, so that the potential-flow analogy will be valid, provided $z/z_0 \ll 1$. In that event

$$\tfrac{1}{2}Kz^2 + C = \phi = -qx$$

Substituting $z = z_1$ when $x = 0$ and eliminating C yields

$$\tfrac{1}{2}K(z_1^2 - z^2) = qx \qquad 5.9$$

Introducing z_0 as defined in equation 5.4 yields again

$$\frac{S_0 x}{z_0} = \tfrac{1}{2}\left(\frac{z_1}{z_0}\right)^2 - \tfrac{1}{2}\left(\frac{z}{z_0}\right)^2 \qquad 5.8$$

Thus, the potential flow analogy is seen to give the same result as a derivation from first principles provided $z \ll z_0$ – a consequence of the small slope.

If the boundary condition $z = z_2$ when $x = L$ is inserted into equation 5.9, the result is

$$\tfrac{1}{2}K(z_1^2 - z_2^2) = qL \qquad 5.10$$

Rearrangement of equation 5.10 gives equation 5.3 and dividing equation 5.9 by equation 5.10 gives equation 5.2. That is, the results derived for a horizontal base may be used as an approximation when the base slopes, provided the base slope is small.

The concept of a streamfunction can be introduced and we have

$$\psi = qy$$

It will be seen that for $z = z_0 =$ constant $\phi = \frac{1}{2}Kz^2 + C$ is constant and $\partial\phi/\partial x = 0$. That is to say, when the analogy is applied to uniform flow over a sloping bed, it forecasts no flow. While this is obviously untrue, it is important in practice in that it facilitates superposition of flow patterns. The base slope S_0 can be regarded as compensating for the resistance to flow. Analysis of the flow system to be superimposed on a uniform flow can then proceed as if the uniform flow were over a horizontal base without resistance.

A more interesting linear system is one where the discharge is not at a constant rate between two reservoirs, but instead, varies with distance – for example, seepage of a uniform rainfall to a water table. For a uniformly distributed inflow

$$q = ax \qquad 5.11$$

Referring to Fig. 5.5 the origin of coordinates is symmetrically placed between two drains, so that $q = 0$ at $x = 0$. Using Dupuit's approximation

$$\frac{ax}{z} = -K\left(\frac{dz}{dx} - S_0\right) \qquad 5.12$$

To solve this differential equation algebraically requires more work than is justified and a graphical technique is recommended, the method of isoclines. Rearranging equation 5.12 yields

$$z = -\frac{\dfrac{ax}{K}}{\dfrac{dz}{dx} - S_0} \qquad 5.13$$

For any particular value of dz/dx, equation 5.13 is a straight line through the origin and with slope $-(a/K)/(dz/dx - S_0)$. Thus, for a set of values of dz/dx, a family of radial lines can be plotted on the xz-plane. These radial lines are the isoclines and any solution curve must cross each isocline with the slope of its tangent given by the appropriate value of dz/dx. If a number of short closely spaced lines are drawn in the appropriate direction on each isocline, a solution curve satisfying the boundary conditions can be sketched quickly and accurately (Fig. 5.6). Note that scale distortion is necessary for accurate construction and this must be taken into account when plotting the isoclines and the direction indications for the solution curve.

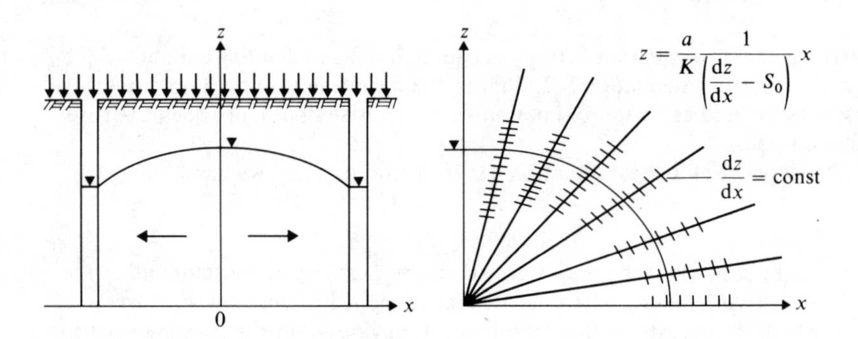

Fig. 5.6. Definition sketch for solution by the isocline method.

If the base slope $S_0 = 0$, an algebraic solution is obtained by shifting the origin to the drain and replacing a by Q/L, where Q is the discharge to one side of the drain. Then, at distance x from the drain

$$q = \frac{L - x}{L} Q$$

That is also given by

$$q = Kz \frac{dz}{dx}$$

using Dupuit's approximation. Elimination of q, followed by integration with the constant given by $z = z_w$ when $x = 0$ yields

$$z^2 - z_W^2 = \frac{2Q}{KL}(Lx - ½x^2)$$

This is the equation of an ellipse and its applicability is discussed at length by van Schilfgaarde *et al.* (1956).

5.4 Radial flow to a well

For radial flow to a well over a horizontal impermeable base, the potential flow analogy yields

$$½Kz^2 + C = \phi = m \ln r \qquad 5.14$$

Substitution of the boundary conditions $z = z_e$ when $r = r_e$ and $z = z_W$ when $r = r_w$ enables m and C to be calculated. The procedure is the same as for flow to a well in a confined aquifer

$$m = \frac{K(z_e^2 - z_w^2)}{2 \ln \frac{r_e}{r_w}} \qquad 5.15$$

$$C = m \ln r_w - ½Kz_w^2 \qquad 5.16$$

If these are used in equation 5.14, the surface profile is given by

$$z^2 - z_w^2 = \frac{z_e^2 - z_w^2}{\ln \frac{r_e}{r_w}} \ln \frac{r}{r_w} \qquad 5.17$$

The flow to the well may be found from the stream function

$$\psi = m\theta \qquad 5.18$$

whence

$$Q = 2\pi m = \frac{\pi K(z_e^2 - z_w^2)}{\ln \frac{r_e}{r_w}} \qquad 5.19$$

Equation 5.19 may be used to evaluate m in terms of Q rather than the boundary water levels. Substitution from equation 5.19 gives

$$z^2 - z_w^2 = \frac{Q}{\pi K} \ln \frac{r}{r_w} \qquad 5.20$$

Base pressures calculated from equation 5.20 agree well with observed piezometric heads; for this reason equation 5.20 is frequently called the base pressure equation.

Although equation 5.3 for linear flow, and equation 5.19 for radial flow, have been derived using the Dupuit approximation, these are in fact exact solutions as was shown by Polubarinova-Kochina (1962), pp. 28–83. (See also Hunt, 1970.)

The relationships between radial velocity and potential and stream function must be remembered – discharge per unit width, not velocity, is given by differentiation. Thus, using equation 5.14

$$q_r = v_r z = -\frac{\partial \phi}{\partial r} = -\frac{m}{r} \qquad 5.21$$

and using equation 5.18

$$q_r = v_r z = -\frac{1}{r}\frac{\partial \phi}{\partial \theta} = -\frac{m}{r}$$

In both cases

$$v_r = -\frac{m}{zr} = -\frac{Q}{2\pi rz}$$

Equation 5.21 can, of course, be taken as a starting point for the derivation of the surface profile.

It is convenient for plotting of flow net and superimposing flows to have $\phi = 0$ when $r = r_w$. This requires a shift from the datum used in equation 5.14 and we take

$$\phi = m \ln \frac{r}{r_w} = \frac{Q}{2\pi} \ln \frac{r}{r_w} = \frac{Q}{4\pi} \ln \frac{x^2 + y^2}{r_w^2} \qquad 5.22$$

The streamlines are given by

$$\psi = m\theta = \frac{Q}{2\pi}\theta = \frac{Q}{2\pi} \tan^{-1}\frac{y}{x} \qquad 5.23$$

Steady radial flow to a well is possible only if there is a source of water to match the flow drawn from the well. If the well is at the centre of an island, as shown in Fig. 3.1, the aquifer is recharged from the lake surrounding the island. When the aquifer is unconfined, recharging by rainfall may occur. Water flows radially in the aquifer and the rate of flow Q is a function of the radius r, Fig. 5.7. If the rainfall rate is P and the flow is steady

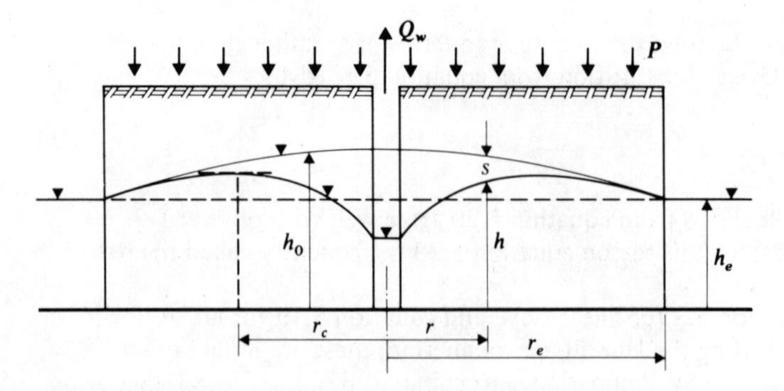

Fig. 5.7. Unconfined radial flow to a well with recharge by rainfall.

$$-dQ = 2\pi r\, P\, dr$$

Hence,

$$-Q = \pi r^2 P + C_1$$

However,

$$Q = -2\pi r\, Kh \frac{dh}{dr}$$

Hence,

$$h\, dh = \frac{\pi r^2 P + C_1}{2\pi K} \frac{dr}{r}$$

and

$$h^2 = \frac{P}{2K} r^2 + \frac{C_1}{\pi K} \ln r + C_2$$

The constants of integration are determined by the following boundary conditions:

$r = r_w \cong 0, \qquad Q = Q_w$

$r = r_e, \qquad h = h_e$

where Q_w is the discharge from the aquifer into the well. Hence,

$$C_1 = -Q_w$$

$$C_2 = h_e^2 - \frac{P}{2K} r_e^2 - \frac{Q}{\pi K} \ln r_e$$

and

$$h_e^2 - h^2 = z_e^2 - z^2 = \frac{P}{2K}(r_e^2 - r^2) + \frac{Q}{\pi K} \ln \frac{r_e}{r}$$

$$Q = Q_w - \pi r^2 P$$

When $Q_w = 0$, the rainfall flows radially outwards through the aquifer and if the height of the water table for this case is h_0, we have

$$h_e^2 - h_0^2 = \frac{P}{2K}(r_e^2 - r^2)$$

The water table for $Q_w > 0$ can then be expressed as

$$h_0^2 - h^2 = \frac{Q}{\pi K} \ln \frac{r_e}{r}$$

Pumping from the well affects the water table for the whole aquifer but only that part within a radius r_c (Fig. 5.7) carries water to the well. The radius r_c is given by

$$\left(\frac{r_c}{r_e}\right)^2 = \frac{Q_w}{P\pi r_e^2}$$

and for an extensive aquifer, this ratio is small. Then, h_0 can be assumed constant, and equal to h_{0c}, say, leading to

$$h_{0c}^2 - h^2 = (h_{0c} - h)(h_{0c} + h) = s(2h_{0c} - s) \simeq 2Hs$$

where $s = h_{0c} - h$ and H is the average depth of water in the aquifer. Hence, the drawdown s can be obtained approximately from

$$s = \frac{Q}{2\pi KH} \ln \frac{r_e}{r}$$

This approximation is valuable because it can be used to estimate the drawdown caused by several wells. It is assumed that the transmissivity KH is the same throughout the aquifer and that linear superposition of the individual drawdowns is valid.

5.5 Well in a uniform flow

If the base of the water bearing soil has a small slope S_0 and the groundwater is free to move, a uniform flow down the slope will be established. To apply the potential flow analogy to this problem, the uniform flow is ignored and the flow to the well analysed as if the base were horizontal and the water at rest, as discussed under linear gravity flow systems. The water surface will then be given by equation 5.17 or equation 5.19, with $z_e = z_0$. If a flow net is desired, this would best be done by plotting the piezometric head h as a function of z and y using

$$h = z \cos \alpha + S_0 x \qquad 5.24$$

The streamlines would be given by a stream function

$$\psi = U_0 z_0 y + \frac{Q}{2\pi} \tan^{-1} \frac{y}{x} \qquad 5.25$$

The components of velocity are obtained by superimposing the components of the elements of the flow as follows:

$$u = -\frac{U_0 z_0}{z} - \frac{Q}{2\pi z} \frac{x}{x^2 + y^2} \qquad 5.26$$

$$v = -\frac{Q}{2\pi z} \frac{y}{x^2 + y^2} \qquad 5.27$$

Here U_0 and z_0 are the velocity and depth for the undisturbed uniform flow.

The flow net is similar to that shown on Fig. 3.5 and the location of the stagnation point $x = x_s, y = 0$ is given by $u = 0$ from equation 5.26

$$x_s = -\frac{Q}{2\pi U_0 z_0} \qquad 5.28$$

The width of the uniform flow diverted to the well is found from equation 5.25 by substituting $x = \infty$ and $\psi = Q/2$; this yields

$$2y = \frac{Q}{U_0 z_0} \qquad 5.29$$

A large number of problems can be handled in the same way. The procedure and the results are similar to those described in more detail for confined flows. The results are satisfactory provided the potential flow analogy and Dupuit's assumptions are valid. The results will not be good near the well, where the slope of the water surface is steep or on a base where the slope is steep.

5.6 Confined flow approximations

A group of approximate solutions to unconfined flow problems were introduced in Chapter 3. It was shown how combinations of sources and sinks can be arranged to represent flow from ponded water or seeping rainfall to drains in a semi-finite or a finite medium. It is also possible to superimpose a uniform flow to deal with flow to drains in a saturated sloping layer, such as a hillside.

However, these are all essentially confined flow problems, where the flow boundaries are known. When the shape of the free surface has to be determined as part of the flow net exact solution becomes much more involved. For these, the reader is referred to the literature of the subject.

5.7 Jaeger's approximation

In the vicinity of a well or a drain the slope of the free surface is much steeper than elsewhere and the assumption of small surface slope is no longer justified. For this type of problem, Jaeger (1956) introduced an alternative approximation in which the equipotential lines are assumed to be parts of circular arcs instead of vertical straight lines. The geometry of these equipotentials is determined by the need for them to be perpendicular to all streamlines, including the free surface and the impermeable base. Jaeger also assumed Darcy's law to be valid.

Fig. 5.8 is a definition sketch for two-dimensional flow to a trench or radial flow to a well. The base is horizontal and the centre of an equipotential is at the intersection of the tangent to the surface and the base. Two equipotentials close together are approximately concentric, so that, for a small increment $\Delta\phi$ in

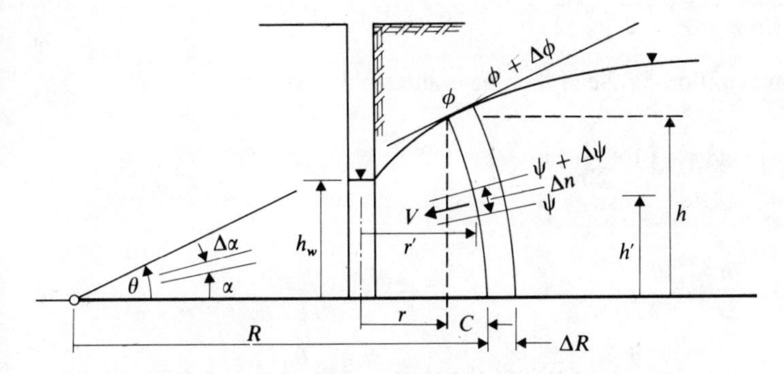

Fig. 5.8. Definition diagram for Jaeger's approximation.

potential, the space between the equipotentials is constant (ΔR) and the velocity

$$V = -\frac{\partial \phi}{\partial s} = -\frac{\partial \phi}{\partial R}$$

The velocity is constant over the arc, the tangents to the streamlines at this location being radial lines. At the free surface by Darcy's law

$$V = -K\frac{\partial h}{\partial R} = -K \sin \theta$$

where θ is the inclination of the free surface. Thus, for flow to a ditch

$$dq = V\,dn = VR\,d\alpha = K \sin \theta \,\frac{h}{\sin \theta}\, d\alpha = Kh\, d\alpha \qquad 5.30$$

and integration yields

$$q = Kh\theta \qquad 5.31$$

Radial flow to a well through the element of area $dA = 2\pi r' dn$, with $0 < \alpha < \theta$ is

$$dQ = 2\pi r' dn\ V = V\, 2\pi r'\, R\, d\alpha$$

From Fig. 5.7, $R = h'/\sin \alpha = h/\sin \theta$, which, on substitution yields

$$dQ = K \sin \theta\ 2\pi r' \,\frac{h}{\sin \theta}\, d\alpha$$

and

$$Q = 2\pi Kh \int_0^\theta r'\, d\alpha \qquad 5.32$$

Referring to Fig. 5.7 again, the mean radius can be expressed as

$$r_m = r + \frac{C}{2} = r + \frac{R}{2}(1 - \cos \theta)$$

$$= r + \frac{h}{2}\,\frac{1 - \cos \theta}{\sin \theta} = r + \frac{h}{2} \tan \frac{\theta}{2}$$

The integral in equation 5.32 can then be evaluated as

$$\int_0^\theta r' d\alpha \doteqdot r_m \int_0^\theta d\alpha = r\left(1 + \frac{h}{2r} \tan \frac{\theta}{2}\right)\theta$$

Hence

$$Q = 2\pi Khr\left(1 + \frac{h}{2r} \tan \frac{\theta}{2}\right)\theta \qquad 5.33$$

For small values of θ, $\tan \frac{\theta}{2}$ is small and the term $\frac{h}{2r} \tan \frac{\theta}{2}$ may be dropped, yielding

$$Q = 2\pi Khr\,\theta \qquad 5.34$$

Equations 5.31, 5.33 and 5.34 are differential equations for the surface profiles in the two cases. However, before proceeding to solve them, the concept of critical depth introduced by Jaeger needs to be investigated.

From equations 5.31, 5.33 and 5.34 it is apparent that the higher the discharge, the steeper the slope of the profile, Fig. 5.9.

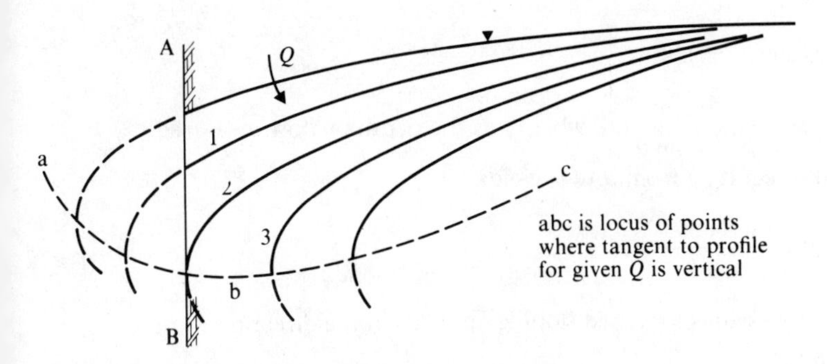

Fig. 5.9. Jaeger's concept of critical depth.

If now the line AB represents the boundary of a well or drain, then it is seen that profile 2 just touches AB whereas profile 3 does not reach AB at all. Hence, with $Q_1 < Q_2 < Q_3$, it follows that Q_2 is the maximum discharge which is physically possible. Lowering of the water level in the well below the point where profile 2 touches the well will not affect the discharge. That is, provided the water level in the well is lower than the intersection of the curve abc and the well face, the discharge to the well will be Q_2 and independent of the water level in the well. This is an important conclusion, for such a phenomenon has been observed, and the Dupuit approximation fails to explain it. If the discharge is less than Q_2, however, the rate of flow does depend on the water level in the well. (In nature there will also be a surface of seepage above the water level.)

In a sense, the outflow water level defined by the curve abc is analogous to critical depth in open channel flow. If $h_w < h_c$, the flow is independent of 'tail water' level.

The critical depth h_c can be calculated from equation 5.31 or 5.33 for a given discharge by substututing $r = r_w$, $h = h_c$ and $\theta = \pi/2$. Hence, for the drain

$$q = Kh_c\,\frac{\pi}{2}$$

$$h_c = \frac{2}{\pi}\,\frac{q}{K} = 0.636\,\frac{q}{K}$$

For the well

$$\frac{Q}{K\pi^2 r_w} = h_c\left(1 + \frac{h_c}{2r_w}\right)$$

and

$$h_c = -r_w + \sqrt{r_w^2 + \frac{2Q}{\pi^2 K}}$$

This method of analysis can also be extended to cases where the impervious base slopes at an angle $\pm S$ (positive upwards). Then

$$V = K \sin(\theta \pm S)$$

and

$$dq = V\, dn = K \sin(\theta \pm S) dn$$

Substituting $dn = R d\alpha = \frac{y}{\sin\theta} d\alpha$, where y is the depth of flow measured vertically and integrating from 0 to θ yields

$$q = K\left(\cos^2 S \pm \frac{\sin S \cos S}{\tan\theta}\right) y\theta$$

where the positive sign is for a bed sloping upwards in the direction of flow. Rearranging gives

$$\theta = \frac{q}{Ky\left(\cos^2 S \pm \frac{\sin S \cos S}{\tan\theta}\right)}$$

and h_c may be calculated as before to give

$$h_c = \frac{q}{K\left(\frac{\pi}{2} \mp S\right)}$$

The concept of the sloping base is also proposed as an acceptable approximation for partially penetrating drains and wells as shown in Fig. 5.10. The partially penetrating well is approximated by the use of an impervious cone of radius r_e under the well. For this well

$$V = K \sin(\theta - S)$$

and

$$dQ = 2\pi K r' \sin(\theta - S) R d\alpha$$

$$= 2\pi K r' y \frac{\sin(\theta - S)}{\sin\theta} \cos S d\alpha$$

$$Q \doteqdot 2\pi K r y \frac{\sin(\theta - S)\cos S}{\sin\theta}\left(1 + \frac{y}{2r} \tan ½(\theta - S)\right)\theta$$

Here, y is the depth of flow measured vertically.

Solution of the differential equations to find surface profiles is most readily achieved by the isocline method. Thus, for the two-dimensional problem

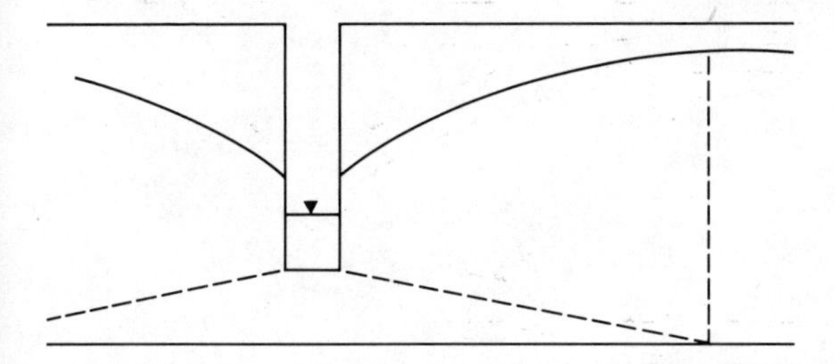

Fig. 5.10. Approximation for a partially penetrating well or drain.

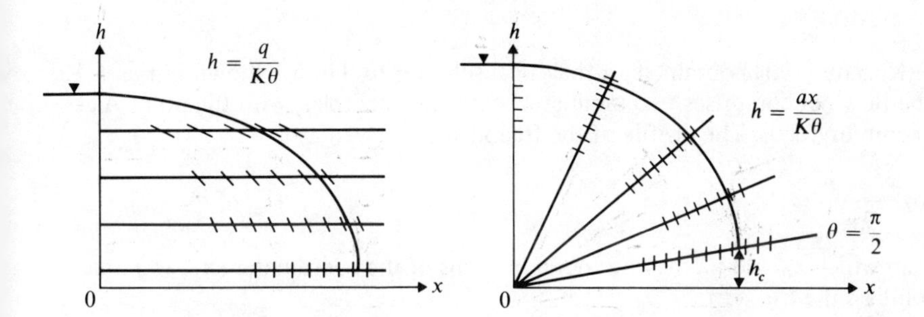

Fig. 5.11. Two-dimensional flow, Jaeger's approximation. Schematic solution by the isocline method.

$$h = \frac{q}{K\theta}$$

For a given discharge and surface slope, the isoclines are horizontal straight lines (Fig. 5.11).
If $q = f(x)$ e.g. $q = ax$, then for a chosen value of θ,

$$h = \frac{ax}{K\theta}$$

The isoclines are a family of straight lines, as discussed under linear gravity flow systems (Fig. 5.5). The working is exactly the same as that described above. It must be remembered that for $\theta = \pi/2$, $h \neq 0$ but $h = h_c$.

5.8 Flow through earth dams

An important group of unconfined gravity flow problems is concerned with seepage through earth dams and stopbanks. Calculation of the shape of the free surface and the length of the surface of seepage is quite complicated and approximate methods, including the use of an electrical analogy and graphical constructions are widely used. These are described in books such as Cedergren (1967) and Sherard *et al.* (1963).

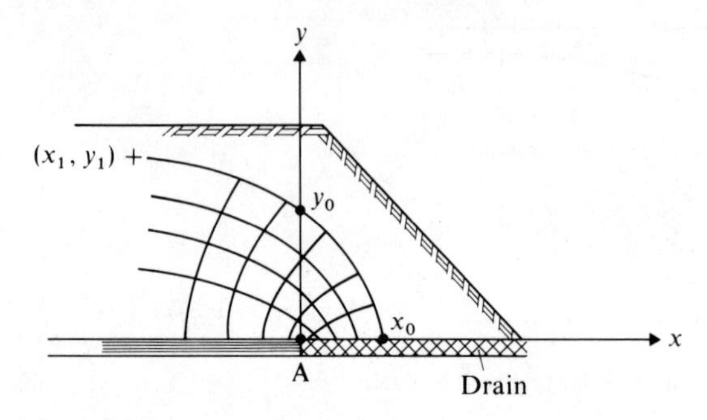

Fig. 5.12. Kozeny solution for flow through earth dam.

Kozeny (1931) obtained an analytical solution to the case shown in Fig. 5.12. The flow net comprises two families of confocal parabolas, with the point A as a common focus. The profile of the free surface is given by

$$x = \frac{y^2 - y_0^2}{2y_0} \qquad 5.35$$

from which x_0 and y_0 may be found in terms of the coordinates (x_1, y_1) of a point on the free surface

$$x_0 = \frac{y_0}{2} = \tfrac{1}{2}\left(\sqrt{x_1^2 + y_1^2} - x_1\right) \qquad 5.36$$

In the case dealt with by Kozeny, there is, of course, no surface of seepage on the sloping face of the dam. Approximate methods, based on Kozeny's analysis, have been derived by Casagrande (1937) and Schaffernak (1917). The toe of the dam is taken to be at A (Fig. 5.12) and the downstream face is inclined at angle β to the horizontal. The Kozeny parabola is constructed as indicated in Fig. 5.13 and, near its apex, it projects through the face of the dam. For $\beta > 60°$, the parabolic surface is assumed to end at the point P where a 45° line from the downstream toe intersects it. The point C is given by the length of the surface of seepage, which is given as a function of β on Fig. 5.13. A fair curve is sketched between P and C.

The flow rate is found by applying Dupuit's approximation to the y axis. There the surface slope is unity, so that

$$q = 2Kx_0 = Ky_0 \qquad 5.37$$

For $\beta < 30°$, the length a is given by Schaffernak

$$a = \frac{x_1}{\cos\beta} - \sqrt{\frac{x_1^2}{\cos^2\beta} - \frac{y_1^2}{\sin^2\beta}} \qquad 5.38$$

and

$$q = Ka \sin\beta \tan\beta \qquad 5.39$$

These formulae are derived by applying Dupuit's approximation at the vertical

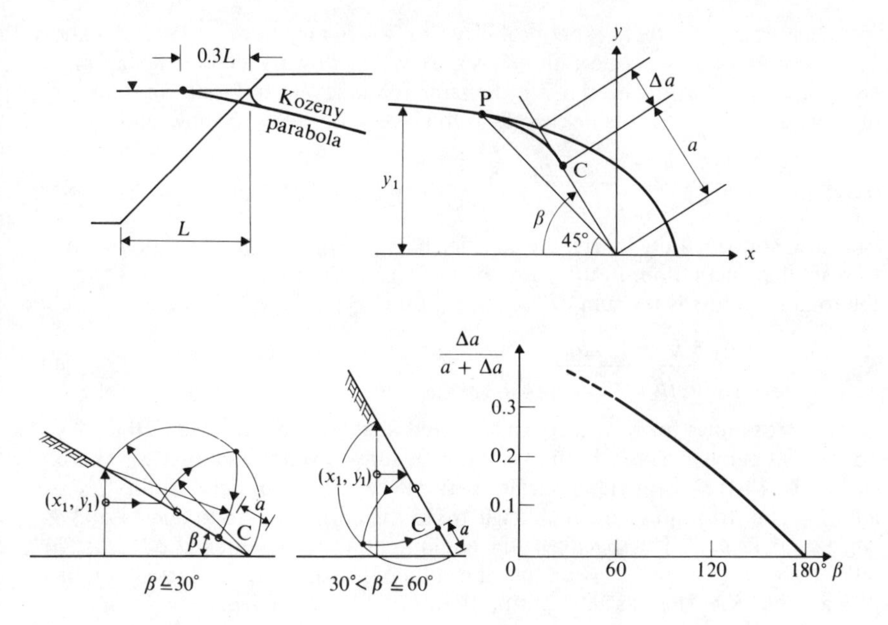

Fig. 5.13. Approximate methods for flow through earth dam based on the Kozeny solution.

through C, assuming the surface slope to be tangential to the downstream face at C.

Equation 5.39 follows immediately from $q = Ky \dfrac{dy}{dx}$

The length a is found by noting that

$$y \frac{dy}{dx} = -a \sin\beta \tan\beta$$

and integrating from the point (x_1, y_1) to C. Then

$$\int y\, dy = -a \sin\beta \tan\beta \int dx$$

$$\tfrac{1}{2}(a^2 \sin^2\beta - y_1^2) = -a \sin\beta \tan\beta\,(-a\cos\beta + x_1) \qquad 5.40$$

When solved for a, equation 5.40 yields equation 5.38.

5.9 Effect of stratification

The homogeneous isotropic medium is an ideal case seldom found in nature, and the coefficient of permeability of most real soils varies with the direction of flow. This may be attributed to the way a medium is formed, layer by layer, as well as consolidation by the weight of the overburden. Natural soils may show distinct layers, and in such soils variations in permeability are obviously to be expected. However, apparently homogeneous soils may have permeability coefficients which depend on direction, and this has to be explained.

The average, or effective, permeability of a layered medium of layer thickness b_i and corresponding permeability K_i is greatest for flow parallel to the layers. Using the facts that the head loss is the same for all layers and that the total discharge is the sum of the discharges in the layers, it is easy to show that

$$K_{max} = \frac{K_1 b_1 + K_2 b_2 + \ldots K_n b_n}{b_1 + b_2 + \ldots b_n} \quad 5.41$$

For flow normal to the layers, the coefficient is a minimum and may be found by a similar calculation. In this case, the velocity is the same for all layers and the total head loss is the sum of the losses in the layers

$$K_{min} = \frac{b_1 + b_2 + \ldots b_n}{b_1/K_1 + b_2/K_2 + \ldots b_n/K_n} \quad 5.42$$

The streamlines for flow through a layered medium are refracted at the boundaries between layers in the same way as surface waves and light waves are. In Fig. 5.14 AB is the dividing surface between two layers of permeability K_1 and K_2. The streamlines are at an angle α_1 to the normal to AB in layer 1 and at angle α_2 in layer 2. The spacing of the equipotentials for an interval $\Delta\phi$ is Δs_1 in layer 1 and Δs_2 in layer 2. Note that the equipotentials must be plotted in terms of h and not Kh. There is a discontinuity in Kh at the boundary, whereas h is continuous.

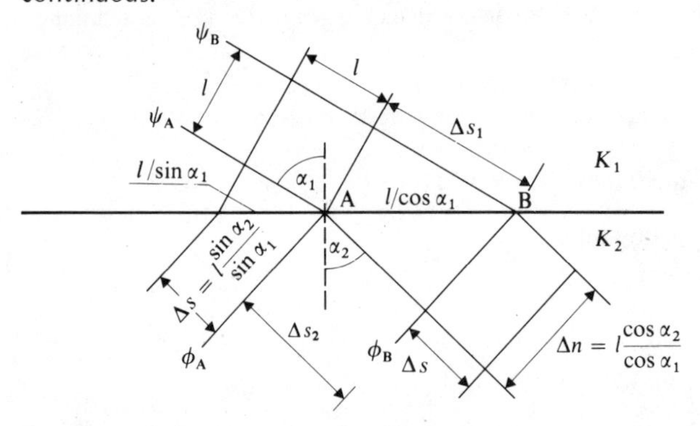

Fig. 5.14. Refraction diagram for flow through a layered medium.

Since the rate of flow between streamlines ψ_A and ψ_B must be the same on both sides of AB

$$K_1 \frac{\Delta h}{\Delta s_1} AB \cos\alpha_1 = K_2 \frac{\Delta h}{\Delta s_2} AB \cos\alpha_2$$

Also

$\Delta s_1 = AB \sin\alpha_1$ and $\Delta s_2 = AB \sin\alpha_2$

so that

$$\frac{K_1}{\tan\alpha_1} = \frac{K_2}{\tan\alpha_2}$$

It is seen that, the lower the permeability, the closer the streamlines are to the normal.

This refraction affects the shape of the flow net. A flow net which is square in layer 1 is rectangular in layer 2. Let two streamlines and two equipotentials from the square flow net in layer 1 be traced across the boundary into layer 2 as shown in Fig. 5.14. The sides of the square in the flow net in layer 1 are all of length l. In layer 2 the equipotentials are spaced $l \sin \alpha_2 / \sin \alpha_1 = \Delta s$ and the streamlines are spaced $l \cos \alpha_2 / \cos \alpha_1 = \Delta n$. Hence

$$\frac{\Delta s}{\Delta n} = \frac{\tan \alpha_2}{\tan \alpha_1} = \frac{K_2}{K_1}$$

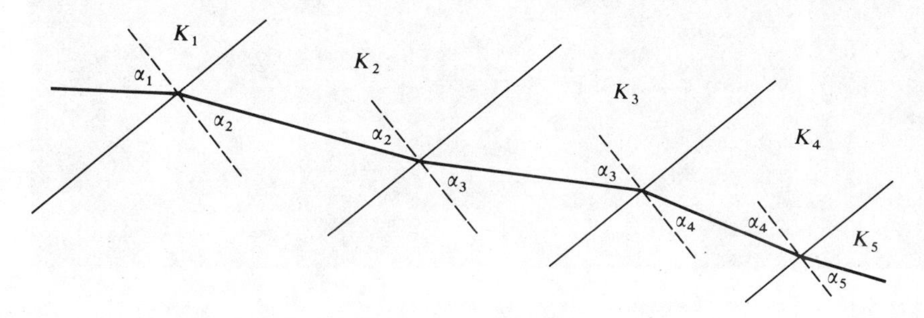

Fig. 5.15. Diagrammatic path of a streamline through layered medium.

The effects of stratification in a multilayered medium can now be seen (Fig. 5.15). The streamlines follow a zigzag course through the layers, their length being shortened through the less permeable layers and lengthened through the more permeable layers by refraction. Thus, refraction distorts the flow net so as to elongate the streamlines parallel to the layers, i.e. in the direction of maximum permeability, and this is borne out by experiment as shown in Fig. 5.16.

These conclusions can be applied to a medium in which the properties vary continuously, by allowing the thickness of the layers to approach zero. Equations 5.41 and 5.42 show that K_{max} and K_{min} depend on the relative thickness of the layers and their relative permeabilities. Provided these relative values do not change as the thickness of each layer tends to zero, K_{max} and K_{min} will be unchanged. Distortion of the flow net will occur, as in the layered system, but the zigzag lines will be replaced by smooth curves. This then, is the case of a soil which is not stratified, but does have a coefficient of permeability which depends on the direction of flow. The most usual case is when the horizontal permeability $K_x > K_z$, the vertical permeability.

Assuming that the components of velocity can be related to the components of grad h by Darcy's law with a different coefficient of permeability for each

$$u = -K_x \frac{\partial h}{\partial x} \quad \text{and} \quad w = -K_z \frac{\partial h}{\partial z}$$

Fig. 5.16. Refraction of streamlines in layered medium.

For a fluid of constant density and steady flow the continuity equation yields

$$K_x \frac{\partial^2 h}{\partial x^2} + K_z \frac{\partial^2 h}{\partial z^2} = 0 \qquad 5.43$$

This can be reduced to the Laplace equation by a scale transformation

$$x_* = x\sqrt{\frac{K_z}{K_x}} = xa$$

Hence,

$$\frac{\partial h}{\partial x} = \frac{\partial h}{\partial x_*}\frac{\partial x_*}{\partial x} = \sqrt{\frac{K_z}{K_x}}\frac{\partial h}{\partial x_*}$$

and

$$\frac{\partial^2 h}{\partial x^2} = \sqrt{\frac{K_z}{K_x}}\frac{\partial}{\partial x}\left(\frac{\partial h}{\partial x_*}\right) = \frac{K_z}{K_x}\frac{\partial^2 h}{\partial x_*^2}$$

Substitution in equation 5.43 yields

$$\frac{\partial^2 h}{\partial x_*^2} + \frac{\partial^2 h}{\partial z^2} = 0$$

Thus, the flow net can be obtained as a solution of the Laplace equation in terms of x_* and z and then replotted to the true scale $x = \sqrt{(K_x/K_z)}\, x_*$.

Physically, the transformed medium is regarded as having constant permeability for all directions equal to K'. Elementary discharges parallel to the two coordinate directions are given by

$$\Delta Q_{x_*} = -K' \Delta z \Delta y \frac{\partial h}{\partial x_*}$$

and

$$\Delta Q_z = -K' \Delta x_* \Delta y \frac{\partial h}{\partial z}$$

For unit gradient of h, the discharge per unit area is the same in each direction. Conversion to true scale requires Δx_* to be replaced by $a\Delta x$ and this affects the gradient of h for flow parallel to the x axis and the area of flow for the component parallel to the z axis. Hence, $a\Delta Q_{x_*} = K'$ dz dy $\partial h/\partial x$ and $\Delta Q_z/a = K'$ dx dy $\partial h/\partial z$ are the corresponding discharges in the real medium. Then the ratio of discharges per unit area for unit gradient in the x and y directions respectively should be equal to the ratio of the permeabilities and since the ratio of these discharges is a^2 this condition is satisfied. Next, the rate of flow has to be calculated from this flow net. In the section on the flat based dam it was shown that $q = (n_s/n_p) KH$. From Fig. 5.17 the discharge between two streamlines in the transfomed medium is

$$\Delta q_t = K' \frac{\Delta h}{\mathrm{AF}} \mathrm{AC} = K' \Delta h$$

since the flow net is made up of squares.

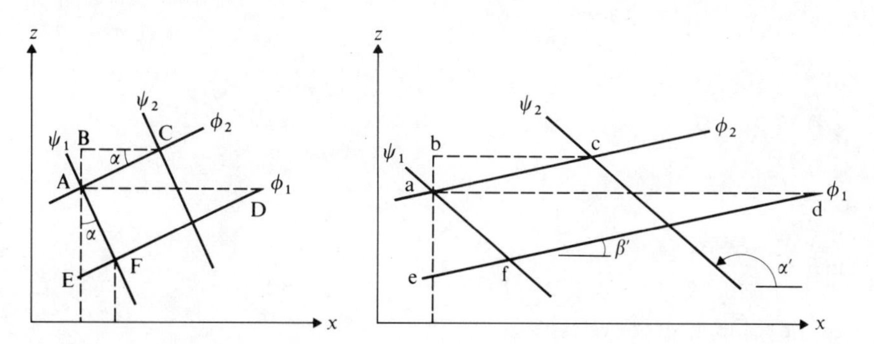

Fig. 5.17. Definition sketch for transformation of the flow net in a medium with $K_x \neq K_y$.

In the real medium, the discharge between the corresponding streamlines is

$$\Delta q = \Delta q_x + \Delta q_z = K_x \frac{\partial h}{\partial x} \mathrm{ab} + K_z \frac{\partial h}{\partial z} \mathrm{bc} \tag{5.44}$$

But

$$\frac{\partial h}{\partial x} = \frac{\Delta h}{\mathrm{ad}} = \frac{\Delta h}{\mathrm{AD}\sqrt{K_x/K_z}} = \frac{\Delta h}{(\mathrm{AF}/\sin \alpha)\sqrt{K_x/K_z}}$$

and

$$\frac{\partial h}{\partial z} = \frac{\Delta h}{\text{ae}} = \frac{\Delta h}{\text{AE}} = \frac{\Delta h}{\text{AF}/\cos \alpha}$$

Furthermore,

$$\text{ab} = \text{AB} = \text{AC} \sin \alpha = \text{AF} \sin \alpha$$

$$\text{bc} = \text{BC} \sqrt{\frac{K_x}{K_z}} \quad = \text{AF} \cos \alpha \sqrt{\frac{K_x}{K_z}}$$

Substituting in equation 5.44 yields, after simplifying

$$\Delta q = \sqrt{K_x K_z}\, \Delta h$$

Now, if the flow through the transformed medium is to represent the flow through the real medium, Δq and Δq_t must be equal and it follows that

$$K' = \sqrt{K_x K_z}$$

The flow rate for the real medium can be found by using the same equation as before, $q = (n_s/n_p)\, KH$, if K' is used instead of K.

It is important to note that the equipotentials and the streamlines are not orthogonal in a homogeneous anisotropic medium. Let α' be the angle between the streamline and $0x$ axis and β' the corresponding angle for an equipotential. Then, for a small change of position along the equipotential, Fig. 5.17,

$$\tan \beta' = \frac{\Delta z}{\Delta x} \quad \text{and} \quad \Delta h = 0$$

but

$$\Delta h = \frac{\partial h}{\partial x} \Delta x + \frac{\partial h}{\partial z} \Delta z$$

so that

$$\tan \beta' = -\frac{\partial h/\partial x}{\partial h/\partial z}$$

For the components of velocity

$$q \cos \alpha' = -K_x \frac{\partial h}{\partial x}$$

$$q \sin \alpha' = -K_z \frac{\partial h}{\partial z}\,.$$

Therefore

$$\frac{\partial h/\partial x}{\partial h/\partial z} = \frac{K_z}{K_x} \frac{1}{\tan \alpha'}$$

and

$$\tan \alpha' \tan \beta' = -\frac{K_z}{K_x}$$

Thus, the flow net is orthogonal only when $K_x = K_z$ or the flow is parallel to one of the principal directions.

The meaning of the permeability of an anisotropic medium in a specified direction s can be investigated as follows. The coefficient of permeability again is defined by a Darcy type of equation as

$$K_s = -\frac{\partial Q/\partial A}{\partial h/\partial s}$$

Choice of the direction s in relation to the flow net is important for, in general, the velocity vector $\mathbf{q}$ and the vector grad h are not parallel. Either might be chosen as the direction s, and the variation of the coefficient with direction can be examined by imagining experiments designed to duplicate flow through a sample in the two cases. The orientation of the samples and the boundary conditions to be simulated are indicated in Fig. 5.18.

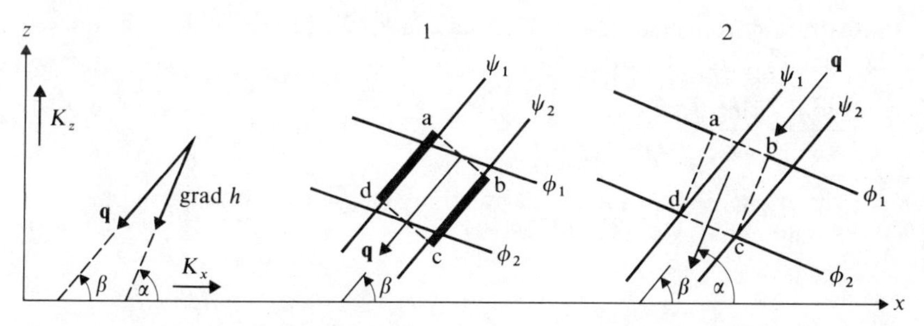

Fig. 5.18. Definition of directional permeability in an isotropic porous medium.

In case 1 the permeability K_{s1} is measured in the direction of $\mathbf{q}$. The specimen is cut from the porous medium at angle β and tested in a permeameter with impervious sides, ad, bc.

$$K_{s1} = -\frac{\partial Q/\partial A}{\partial h/\partial s} = -\frac{|q|}{\Delta h/\Delta s} \qquad 5.45$$

In case 2, the sample is cut in the direction of grad h, i.e. perpendicular to the equipotential lines. It is tested in a permeameter with porous walls ad and bc to simulate its boundary conditions *in situ* and the potential is constant over the ends ab and cd.

$$K_{s2} = -\frac{\partial Q/\partial A}{\partial h/\partial s} = -\frac{\Delta Q/\Delta A}{|\text{grad } h|} \qquad 5.46$$

From equation 5.45

$$\frac{\Delta h}{\Delta s} = -\frac{|q|}{K_{s1}} \qquad 5.47$$

The components of **q** are given by

$$u = |q| \cos\beta = -K_x \frac{\partial h}{\partial x}$$

$$w = |q| \sin\beta = -K_z \frac{\partial h}{\partial z}$$

whence

$$\frac{\partial h}{\partial x} = -\frac{|q| \cos\beta}{K_x} \qquad 5.48$$

$$\frac{\partial h}{\partial z} = -\frac{|q| \sin\beta}{K_z} \qquad 5.49$$

But

$$\frac{\Delta h}{\Delta s} = \frac{\partial h}{\partial x}\frac{\Delta x}{\Delta s} + \frac{\partial h}{\partial z}\frac{\Delta z}{\Delta s}$$

and substituting from equations 5.47, 5.48 and 5.49 with $\Delta x/\Delta s = \cos\beta$ and $\Delta z/\Delta s = \sin\beta$ yields in equation 5.45

$$\frac{1}{K_{s1}} = \frac{\cos^2\beta}{K_x} + \frac{\sin^2\beta}{K_z} \qquad 5.50$$

$\sqrt{K_{s1}}$ plotted against β on a polar diagram is an ellipse with semimajor axis $\sqrt{K_x}$ and semiminor axis $\sqrt{K_z}$ (Fig. 5.19).

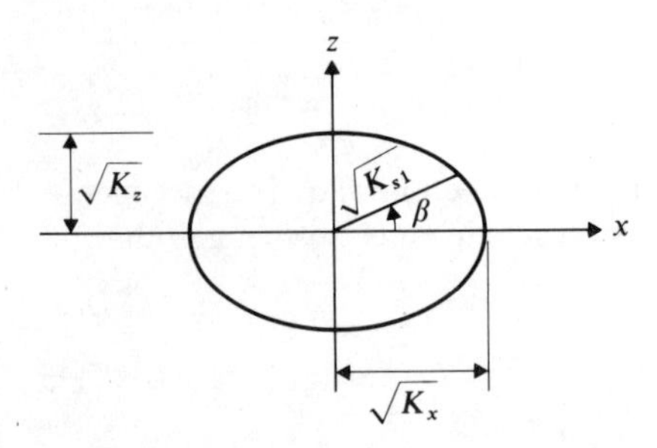

Fig. 5.19. Directional permeability K_{s1}.

For case 2

$$K_{s2} = -\frac{\Delta Q/\Delta A}{|\operatorname{grad} h|} \qquad 5.46$$

and

$$\frac{\Delta Q}{\Delta A} = q\cos(\alpha - \beta) \qquad 5.51$$

$$u = |q| \cos\beta = -K_x \frac{\partial h}{\partial x} \qquad 5.52$$

$$w = |q| \sin\beta = -K_z \frac{\partial h}{\partial z} \qquad 5.53$$

But

$$\frac{\partial h}{\partial x} = |\text{grad } h| \cos\alpha \qquad 5.54$$

$$\frac{\partial h}{\partial z} = |\text{grad } h| \sin\alpha \qquad 5.55$$

From equations 5.46 and 5.51

$$|q| = -\frac{K_{s2}|\text{grad } h|}{\cos(\alpha - \beta)} \qquad 5.56$$

so that, using equations 5.52 to 5.56

$$K_x \cos\alpha = \frac{K_{s2} \cos\beta}{\cos(\alpha - \beta)} \qquad 5.57$$

and

$$K_z \sin\alpha = \frac{K_{s2} \sin\beta}{\cos(\alpha - \beta)} \qquad 5.58$$

When equation 5.57 is multiplied by $\cos\alpha$ and equation 5.58 by $\sin\alpha$ and the results added,

$$K_{s2} = K_x \cos^2\alpha + K_z \sin^2\alpha \qquad 5.59$$

The plot of $1/\sqrt{K_{s2}}$ against α on a polar diagram is also an ellipse, with semimajor axis $1/\sqrt{K_z}$ and semiminor axis $1/\sqrt{K_x}$, Fig. 5.20. In particular, when $\alpha = 0$ or $90°$, $\beta = 0$ or $90°$ also. The velocity and grad h vectors are parallel and both experiments would yield the same result.

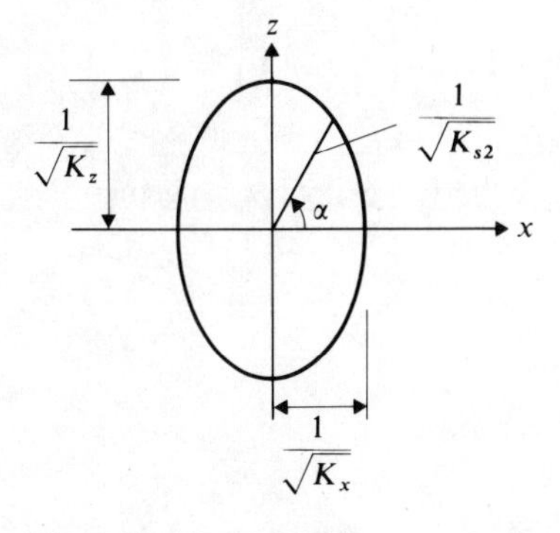

Fig. 5.20. Directional permeability K_{s2}.

The difference between the values of K_s for a given direction θ can be obtained from the ratio K_{s2}/K_{s1}, obtained by multiplying equations 5.50 and 5.59

$$\frac{K_{s2}}{K_{s1}} = 1 + \left(\frac{K_z}{K_x} + \frac{K_x}{K_z} - 2\right) \sin^2\theta \cos^2\theta$$

The difference is seen to vanish when $\theta = 0$ or $90°$ or when $K_x = K_z$. It is a maximum when $\theta = 45°$. The difference may be illustrated by superimposing a plot of $\sqrt{K_{s2}}$ versus θ on the ellipse in Fig. 5.19. This has been done in Fig. 5.21, where the lobe shaped diagram gives $\sqrt{K_{s2}}$ and the ellipse $\sqrt{K_{s1}}$.

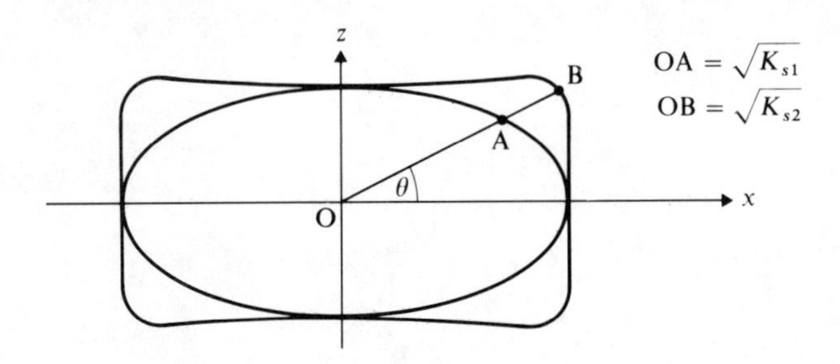

Fig. 5.21. K_{s1}, K_{s2} as functions of θ.

Example 5.1

The shorelines of two lakes are approximately straight and parallel and 2300 m apart. The ground between them comprises permeable soil ($K = 3.8 \times 10^{-4}$ m/s) overlying an impervious bed which has a slope of 1 in 3400. The depth of water over the impermeable layer is 1.8 m at the shoreline of the higher lake and it is 3.2 m at the other shoreline.

Calculate the flow rate per unit length of shoreline.

Since $z_2 > z_1$, the flow is diverging and $z_1 > z_0$ also. Hence, equation 5.6 is relevant and

$$\frac{S_0 L}{z_0} = \left[\left(\frac{z_2}{z_0} - 1\right) - \ln\left(\frac{z_2}{z_0} - 1\right)\right] - \left[\left(\frac{z_1}{z_0} - 1\right) - \ln\left(\frac{z_1}{z_0} - 1\right)\right]$$

where L is the distance between the lakes.

Since z_0 is the only unknown in the equation, the following rearrangement is made as a preliminary to solving for z_0:

$$S_0 L = (z_2 - z_1) + z_0 \ln\frac{z_1 - z_0}{z_2 - z_0}$$

$$z_0 \ln\frac{z_2 - z_0}{z_1 - z_0} = (z_2 - z_1) - S_0 L$$

$z_1 = 1.8$ m, $z_2 = 3.2$ m, $S_0 = 1/3400$, $L = 2300$ m

Hence,

$$z_0 \ln \frac{3.2 - z_0}{1.8 - z_0} = 0.723$$

By trial and error:

z_0	$= 0.8$	0.9	0.82
$3.2 - z_0$	$= 2.4$	2.3	2.38
$1.8 - z_0$	$= 1.0$	0.9	0.98
$z_0 \ln \frac{3.2 - z_0}{1.8 - z_0}$	$= 0.699$	0.842	0.725

$$z_0 = 0.82 \text{ m}$$

From equation 5.4

$$q = KS_0 z_0$$

$$= 3.8 \times 10^{-4} \times \frac{1}{3400} \times 0.82$$

$$= 9.16 \times 10^{-8} \text{ m}^2/\text{s}$$

Example 5.2

Water is pumped at the rate of 1300 l/hour from a well in an unconfined aquifer where the permeability K is 4.3×10^{-5} m/s. The diameter of the well is 200 mm.

Calculate the critical depth for this flow. Sketch the water table when the depth in the well equals critical depth using Jaeger's method.

Critical depth h_c is given by equation 5.33 with $\theta = \pi/2$ and $r = r_w$

$$\frac{Q}{K\pi^2 r_w} = h_c \left(1 + \frac{h_c}{2r_w}\right)$$

Hence,

$$h_c = -r_w + \sqrt{r_w^2 + \frac{2Q}{\pi^2 K}}$$

$$\frac{2Q}{\pi^2 K} = \frac{2 \times 1.3}{3600 \times \pi \times \pi \times 4.3 \times 10^{-5}} = 1.702$$

$$h_c = -0.1 + \sqrt{0.01 + 1.702}$$

$$= 1.208 \text{ m}$$

The water table is given by equation 5.33

$$\frac{Q}{2\pi K} = hr \left(1 + \frac{h}{2r} \tan \frac{\theta}{2}\right) \theta$$

$$\frac{Q}{2\pi K} = 1.3366$$

$$\theta hr + \frac{h^2}{2}\theta \tan\frac{\theta}{2} = 1.3366$$

The isoclines are given by

$$r = \left(1.3366 - \frac{h^2}{2}\theta \tan\frac{\theta}{2}\right)/(\theta h)$$

$$= 1.3366/(\theta h) \text{ (for } \theta \to 0)$$

θ	Isocline
80°	$r = (1.3366 - 5.8580 \times 10^{-1} h^2)/1.3963\, h$
70°	$r = (1.3366 - 4.2773 \times 10^{-1} h^2)/1.2217\, h$
60°	$r = (1.3366 - 3.0230 \times 10^{-1} h^2)/1.0472\, h$
50°	$r = (1.3366 - 2.0347 \times 10^{-1} h^2)/0.87266\, h$
40°	$r = (1.3366 - 1.2705 \times 10^{-1} h^2)/0.69813\, h$
30°	$r = (1.3366 - 7.0149 \times 10^{-2} h^2)/0.52360\, h$
20°	$r = (1.3366 - 3.0775 \times 10^{-2} h^2)/0.34907\, h$
10°	$r = (1.3366 - 7.6348 \times 10^{-3} h^2)/0.17453\, h$
6°	$r = (1.3366 - 2.7441 \times 10^{-3} h^2)/0.10472\, h$
4°	$r = (1.3366 - 1.2190 \times 10^{-3} h^2)/0.069813\, h$
2°	$r = (1.3366 - 3.0465 \times 10^{-4} h^2)/0.034907\, h$

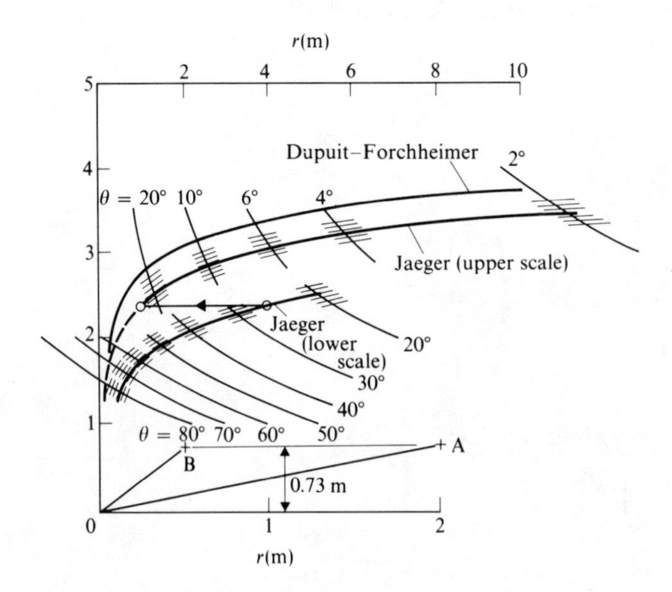

Fig. 5.22. Graphical solution for example 5.2 with solution for example 5.3 superimposed.

These isoclines are drawn on Fig. 5.22 and short lengths of tangents are drawn in the appropriate directions and intersecting the isoclines. Note, the different scales in the vertical and horizontal directions. The direction of tangents are best found by using $\tan\theta$; for example, when $\theta = 20^\circ$, $\tan\theta = 0.3640$ corresponding to a rise of 0.73 m in 2 m. Thus, for the tangents crossing the 20° isocline 0A gives the direction when the lower horizontal scale is used and 0B when the upper horizontal scale is used.

Example 5.3

For the same flow as in Example 5.2, determine the water table using the Dupuit-Forchheimer assumption and sketch the curve,

From equation 5.20

$$z^2 - z_w^2 = \frac{Q}{\pi K} \ln \frac{r}{r_w}$$

$$\frac{Q}{\pi K} = 2.6732$$

$$z_w^2 = h_c^2 = 1.4593 \text{ m}^2$$

$$r_w = 0.1 \text{ m}$$

Hence,

$$z^2 = 1.4593 + 2.6732 \ln \frac{r}{0.1}$$

$r(m)$	0.2	0.4	0.6	0.8	1.0
$z(m)$	1.82	2.27	2.50	2.65	2.76

$r(m)$	2	3	4	5	6	7	8	9	10
$z(m)$	3.08	3.25	3.36	3.45	3.52	3.58	3.63	3.67	3.71

These values of r and $z(= h)$ are plotted on Fig. 5.22 for comparison with the results obtained using Jaeger's method.

6 Unsteady groundwater flow

A significant group of groundwater problems is concerned with unsteady flows. The variation with time may be transient, as it is during the establishment of a steady flow, or it may be long lasting; for example, when water is pumped from a reservoir with no inflow. In this section unsteady flow through an aquifer to a well will be discussed. An important application of the results of the analysis is the determination of the hydraulic properties of an aquifer by pumping tests.

In Chapter 2 the principle of conservation of matter and Darcy's law were used to derive equation 2.18, the continuity equation for unsteady flow:

$$\nabla^2 h = \frac{\beta\rho g\left(n + \dfrac{\alpha}{\beta}\right)}{K}\frac{\partial h}{\partial t} \qquad 2.18$$

We recall from Chapter 2 the definitions of n, α and β and that $\rho g\alpha$ is the water drawn from storage because of compression of the aquifer while $\beta\rho g n$ comes from expansion of the water. The factor $\beta\rho g(n + \alpha/\beta)$ is the specific storage S_s and it is convenient to define a dimensionless storage coefficient

$$S = bS_s$$

where b is the thickness of the aquifer.

The transmissivity of the aquifer can be defined as

$$T = bK$$

and using these equation 2.18 becomes

$$\nabla^2 h = \frac{S}{T}\frac{\partial h}{\partial t} \qquad 6.1$$

Analytical investigation of unsteady groundwater flow was first undertaken by Theis (1935), who solved the problem of radial flow to a fully penetrating well in an extensive confined aquifer. There is no inflow and the entire discharge comes from water in storage. The piezometric surface varies with time and in an interval of time Δt, the volume of water released from an element of the aquifer at (r, θ) is

$$dQ\Delta t = -Sr\, d\theta\, dr \frac{\partial h}{\partial t} \Delta t$$

where $Q(t)$ is the rate of flow to the well.
Hence

$$Q(t) = -S \int_{r_w}^{\infty} \int_{0}^{2\pi} r \frac{\partial h}{\partial t} d\theta\, dr \qquad 6.2$$

In general

$$h = f(r, \theta, t)$$

However, for a single well in an infinitely large homogeneous aquifer there is radial symmetry and equation 6.2 becomes

$$Q(t) = -2\pi S \int_{r_w}^{\infty} r \frac{\partial}{\partial t} h(r, t) dr \qquad 6.3$$

Since there is no inflow, there is no steady state solution and the area affected by pumping increases with time.

Theis transformed equation 6.1 into polar coordinates,

$$\frac{\partial^2 h}{\partial r^2} + \frac{1}{r}\frac{\partial h}{\partial r} = \frac{S}{T}\frac{\partial h}{\partial t}$$

and to solve it used the known solution to the analogous problem of conduction of heat to a sink of constant strength. The boundary conditions to be satisfied are as follows:

$$h = h_0 \quad \text{as} \quad r \to \infty \quad \text{for} \quad t > 0$$

$$\lim_{r \to 0} \left(r \frac{\partial h}{\partial r} \right) = \frac{Q}{2\pi T}$$

$$t \leqslant 0, h(r, o) = h_0$$

Note that the second of these expresses the fact that, at the face of the well

$$Q = \left(K \frac{\partial h}{\partial r} \right) (2\pi r b)$$

while the third condition says that the piezometric surface is horizontal and the water in the aquifer is at rest before pumping begins.

If the rate of pumping, Q, is constant, substitution will verify that the drawdown is given by

$$\Delta h = h_0 - h = \frac{Q}{4\pi T} \int_{\frac{r^2 S}{4Tt}}^{\infty} \frac{1}{x} e^{-x}\, dx \qquad 6.4(a)$$

$$= \frac{Q}{2\pi T} \int_{\frac{r}{\sqrt{4t(T/S)}}}^{\infty} \frac{1}{u} e^{-u^2} \, du \qquad 6.4(b)$$

The integrals in equations 6.4 (a) and (b) are available in one form or the other as functions of the lower limit in tables in Jahnke and Emde (1945), Wenzel and Fishel (1942), Glover (1966) and Cedergren (1967). If the integral in equation 6.4 (a) is defined by

$$-Ei(-m) = \int_{m}^{\infty} \frac{1}{x} e^{-x} \, dx$$

then, in equation 6.4(b)

$$\int_{\sqrt{m}}^{\infty} \frac{1}{u} e^{-u^2} \, du = -\tfrac{1}{2} Ei\,(-m) \equiv W(u^2)$$

In either case, Δh is found by putting

$$m = \frac{r^2 S}{4Tt} \qquad 6.5$$

If Q is not constant, the linearity of equation 6.1 can be used to superimpose solutions. The varying discharge can be represented, at least approximately, by a succession of constant flow rates and the computed values of Δh can be added. Note that, for each solution, $t = 0$ is at the start of pumping.

The Theis equation for drawdown depends on many assumptions, few of which are satisfied exactly in the field; for example, complete penetration of the well, infinite extent in area of the aquifer which is homogeneous and isotropic, infinitesimal diameter of the well. Nevertheless, it is widely and successfully used to predict drawdown and flow rate and to determine the properties T and S of the aquifer.

For the latter purpose, equation 6.4(a) is rewritten as

$$\Delta h = \frac{Q}{4\pi T} I \qquad 6.6$$

where

$$I = -Ei\,(-m)$$

Taking the logarithm to base 10 of both sides of equation 6.6

$$\log \Delta h = \log \frac{Q}{4\pi T} + \log I \qquad 6.7$$

Also, from equation 6.5

$$\log \frac{r^2}{t} = \log \frac{4T}{S} + \log m \qquad 6.8$$

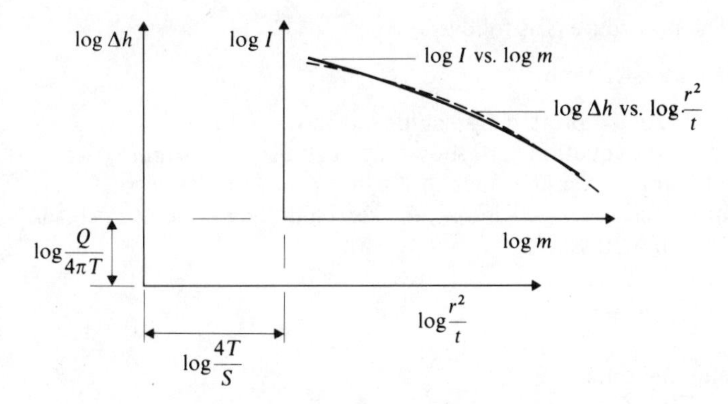

Fig. 6.1. Graphical method of solution of equations for the formation constants T and S.

Now I is a known function of m and Δh can be found as a function of r^2/t by means of a pumping test – water is pumped at constant rate from a well and Δh, r and t are measured at one or more observation wells. Equations 6.7 and 6.8 show that the coordinates of the two functions are related to each other by translation of their axes without rotation, as in Fig. 6.1.

This is the basis of a graphical method for finding T and S from the results of a pumping test, Theiss (1935), De Wiest (1965). Using tables I is plotted against m on log-log scales. Using the results of the pumping test, Δh is plotted against r^2/t, also on log-log scales of the same size. If one of these curves is on transparent paper, it can be made to fit over the other, the axes of the curves remaining parallel to each other as in Fig. 6.1. The shift of origin gives $\log Q/4\pi T$ and $\log 4T/S$ and hence T and S. In practice, T and S may be found most easily by reading I, m, Δh and r^2/t from a convenient point common to both fields. Substitution in equations 6.5 and 6.6 leads directly to the desired result.

Jacob (1950) has suggested a method of finding the properties T and S based on expansion of the integrand in equation 6.4(a). When the expansion is integrated, the result is a convergent series and equation 6.4(a) becomes

$$\Delta h = \frac{Q}{4\pi T}\left[-0.5772 - \ln m + m - \frac{m^2}{2.2!} + \frac{m^3}{3.3!} - \frac{m^4}{4.4!} + - - -\right] \qquad 6.9$$

Provided m is small enough, say less then 0.01, the series may be terminated in the second term. Thus, for small values of r and large values of t, equation 6.9 is approximately the same as

$$\Delta h = \frac{Q}{4\pi T}(-\ln 1.78 - \ln m)$$

$$= 2.30\frac{Q}{4\pi T}\log\frac{2.25\,Tt}{r^2 S} \qquad 6.10$$

Equation 6.10 may be applied to observations made in one well at successive times, to several wells and one time or to several wells and different times. Numerical examples are given by Cooper and Jacob (1946), Jacob (1950) and

De Wiest (1965). The procedure is as follows

(i) Single well and successive times.

The observed values of Δh are plotted against log t as shown on Fig. 6.2(a). Provided t is large enough, equation 6.10 shows that a straight line should pass through the plotted points. From the slope of the line T may be obtained. To find S, the line is projected to $\Delta h = 0$ and t_0, the corresponding value of t is read. Substitution in equation 6.10 yields

$$S = 2.25 \frac{Tt_0}{r^2}$$

(ii) Several wells and one time.

In this case, Δh is plotted against log r as shown in Fig. 6.2(b). From equation 6.10 it is clear that T may be obtained from the slope of the straight line which fits the data for small values of r, while S is given by

$$S = \frac{2.25\ Tt}{r_0^2}$$

The radius r_0 is found by extending the straight line to $\Delta h = 0$.

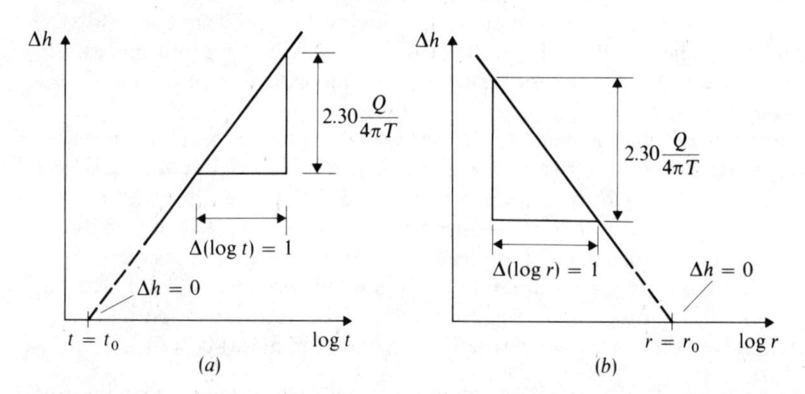

Fig. 6.2. Determination of formation constants from pumping test data. (a) Single well. (b) Several wells.

(iii) Several wells and different times.

When equation 6.10 is written in the form

$$\Delta h = 2.30 \frac{Q}{4\pi T} \log \frac{2.25T}{S} - 2.30 \frac{Q}{4\pi t} \log \frac{r^2}{t}$$

it is clear that the same technique can be employed by plotting Δh against log r^2/t and fitting a straight line for small values of that variable.

The formation constants T and S can also be obtained from recovery data obtained after pumping at a constant rate has ceased. The water level in the well will rise from h_{t_0} at time t_0 when pumping ceased to h_t at some later time t. The drawdown at that time is $h_0 - h_t$, h_0 being the elevation of the piezometric surface before pumping started. Equation 6.1 being linear, $h_0 - h_t$ can be calculated by assuming that pumping continues after time t_0 and superimposing the solution for a discharge of $-Q$ beginning at t_0. The radius of the well r_w is substituted for r and the result is

$$h_0 - h_t = \frac{Q}{4\pi T}\left[\int_{\frac{r_w^2 S}{4Tt}}^{\infty} \frac{1}{x} e^{-x}\,dx - \int_{\frac{r_w^2 S}{4T(t-t_0)}}^{\infty} \frac{1}{x} e^{-x}\,dx\right] \qquad 6.11$$

The technique of graphical superposition described earlier may be applied by first rewriting equation 6.11 in the form

$$h_0 - h_t = \frac{Q}{4\pi T}[I_1 - I_2]$$

where

$$I_1 = -Ei(-m_1) \qquad m_1 = \frac{r_w^2 S}{4Tt}$$

$$I_2 = -Ei(-m_2) \qquad m_2 = \frac{r_w^2 S}{4T(t - t_0)}$$

$$t > t_0$$

Note that I_1 and I_2 are not independent since

$$\frac{1}{m_2} = \frac{1}{m_1} - \frac{4Tt_0}{r_w^2 S}$$

and $\dfrac{4Tt_0}{r_w^2 S}$ is constant.

Hence

$$\log(h_0 - h_t) = \log\frac{Q}{4\pi T} + \log(I_1 - I_2)$$

and

$$\log t = \log\frac{r_w^2 S}{4T} + \log\frac{1}{m_1}$$

Plotting $(h_0 - h_t)$ against t and $(I_1 - I_2)$ against $1/m_1$, both on log-log scales enables T and S to be calculated as before.

The approximation used by Jacob can also be used here. Thus, for any value of r, including $r = r_w$, provided $(r^2S)/(4Tt)$ is small

$$h_0 - h_t = 2.30\frac{Q}{4\pi T}\left[\log\frac{2.25Tt}{r^2S} - \log\frac{2.25T(t - t_0)}{r^2S}\right]$$

$$= 2.30\frac{Q}{4\pi T}\log\frac{t}{t - t_0} \qquad 6.12$$

By fitting a straight line to $(h_0 - h_t)$ plotted against $\log\left(\frac{t}{t - t_0}\right)$, T and S can be calculated.

As in the study of steady flows, the linear character of equation 6.1 permits superposition of the effects of several wells, including the use of images to satisfy boundary conditions. For example, unsteady flow in a confined aquifer with lateral inflow from a river or lake is solved by using an image of strength $-Q$. The drawdown is then

$$\Delta h = \frac{Q}{4\pi T}\left[-Ei\left(-\frac{r_1^2 S}{4Tt}\right)\right] - \frac{Q}{4\pi T}\left[-Ei\left(\frac{r_2^2 S}{4Tt}\right)\right] \qquad 6.13$$

After a suitably long time, Jacob's approximation is valid and equation 6.13 is closely represented by

$$\Delta h = \frac{Q}{4\pi T}\left[ln\,\frac{4Tt}{1.78 r_1^2 S} - ln\,\frac{4Tt}{1.78 r_2^2 S}\right]$$

$$= \frac{Q}{4\pi T}\,ln\,\frac{r_2^2}{r_1^2} \qquad 6.14$$

which is the same as equation 3.34 obtained for steady flow. This is, of course, to be expected since the unsteadiness of this flow is transient and the condition for the approximation to be valid $\left(\frac{r^2 S}{4Tt}\ \text{small}\right)$ is satisfied for t large. Such a system tends towards a steady flow from the river or the lake to the well as time passes.

Equation 6.1 can also be used to relate groundwater fluctuations to river level fluctuations when the confined aquifer is linked with the river. It may also be applied to unconfined aquifers where the variations in thickness are small. Tison (1965) showed that if the river depth varies as

$$\Delta h(0,t) = a_0 \sin \frac{2\pi}{\tau}\, t$$

where Δh is the departure from mean level, a_0 is the amplitude of a sine wave with period τ, then with the x axis normal to the river the solution of equation 6.1 is given by

$$\Delta h\,(x, t) = a_0 \exp\left[-x\sqrt{\frac{\pi S}{\tau K b}}\right] \sin\left(\frac{2\pi t}{\tau} - x\sqrt{\frac{\pi S}{\tau K b}}\right)$$

The level fluctuations within the aquifer decrease with distance from the river. The amplitude at a distance x from the river is

$$\Delta h_0 = a_0 \exp\left(-x\sqrt{\frac{\pi S}{\tau K b}}\right)$$

or

$$\log \Delta h_0 = \log a_0 - x\sqrt{\frac{\pi S}{\pi K b}}\,\log e$$

and the waves travel in the aquifer with the velocity

$$V = \frac{2\pi/\tau}{\sqrt{\pi S/\tau Kb}} = \sqrt{\frac{4\pi Kb}{\tau S}}$$

Hence, S/Kb can be obtained from well observations at known distance from the river or an ocean; level fluctuations as a function of time.

A very interesting discussion of the flow from an aquifer with recharge is given by Hammad (1969) who uses the unsteady flow equation to predict the yield of artesian wells in the oases of the eastern Sahara. A line of equally spaced wells is postulated, and the line is normal to the direction of the undisturbed flow in the aquifer. Close to the wells, the streamlines are curved, as the water enters the wells, but apart from this, the flow is approximately two dimensional. Equation 6.1 becomes

$$\frac{\partial^2 h}{\partial x^2} = \frac{S}{T}\frac{\partial h}{\partial t}$$

The problem is solved for a given water level in the wells, h_0. For an artesian aquifer and no pumping, this will be ground level in the oasis. The solution is found by superposition of two flows. They are a steady flow along the aquifer due to the slope S_0 of the piezometric surface and an unsteady flow from storage, with the undisturbed piezometric surface horizontal.

For the steady flow, the height of the piezometric surface above water level in the well is

$$h_1 - h_0 = S_0 x \qquad 6.15$$

and this part of the solution satisfies

$$\frac{\partial^2 (h_1 - h_0)}{\partial x^2} = \frac{S}{T}\frac{\partial (h_1 - h_0)}{\partial t} = 0$$

For the unsteady flow we have

$$\frac{\partial^2 (h_2 - h_0)}{\partial x^2} = \frac{S}{T}\frac{\partial (h_2 - h_0)}{\partial t}$$

and the following boundary conditions:

$t = 0$ and $0 < x < \infty$ $\quad h_2 - h_0 = H_0$

$0 < t < \infty$ and $x = 0$ $\quad h_2 - h_0 = 0$

$0 < t < \infty$ and $x = \infty$ $\quad h_2 - h_0 = H_0$

The solution is

$$h_2 - h_0 = \frac{2H_0}{\sqrt{\pi}} \int_0^{\frac{x}{\sqrt{4t(T/S)}}} e^{-u^2}\, du \qquad 6.16$$

Superimposing these gives $h - h_0$, the elevation of the piezometric surface above water level in the wells

$$h - h_0 = S_0 x + \frac{2H_0}{\sqrt{\pi}} \int_0^{\frac{x}{\sqrt{4t(T/S)}}} e^{-u^2} \, du \qquad 6.17$$

The flow rate per unit length of the line of wells is found by using Darcy's law.

$$q = T \left(\frac{\partial h}{\partial x} \right)_{x=0}$$

$$= T \left(S_0 + \frac{2H_0}{\sqrt{\pi}} \frac{1}{\sqrt{4t(T/S)}} e^{-\frac{x^2}{4t(T/S)}} \right)_{x=0}$$

$$q = TS_0 + H_0 \sqrt{\frac{TS}{\pi}} \frac{1}{\sqrt{t}} \qquad 6.18$$

The first term, TS_0, is the steady flow from the remote source and the second is the unsteady flow from storage in the aquifer. Since T, S, S_0 and H_0 are constants, equation 6.18 is of the form

$$q = A + \frac{B}{\sqrt{t}} \qquad 6.19$$

Although the formation constants could be used to evaluate A and B the practical purpose of estimating the steady flow available at the oasis can be achieved by plotting observed values of q against $1/\sqrt{t}$. A straight line which fits the observations will lead to values of A and B.

To conclude this chapter, we draw attention to two other matters. One is that solutions for flow in a confined aquifer have been applied successfully to unconfined flows when the drawdown is small compared with the initial depth of water in the aquifer. Such flows are examined in more detail in Chapter 8.

The second matter concerns pumping rates during tests. These are not constant if the discharge from the well is unthrottled because the lift increases with time as the water level in the well descends. In more refined analyses, this can be allowed for; Mahmond and Scott (1963), Mahmond *et al.* (1964), Aron and Scott (1965).

In general, unsteady groundwater flow is a complex field of study and only the basic ideas have been introduced here. For more extensive analysis, reference should be made to the literature. (Kraijenhoff van de Leur (1958), De Wiest (1960), Hantush (1959a, 1961, 1962a, 1962c).)

Example 6.1

Water is pumped from a confined aquifer at the rate of 7 l/s. The transmissivity of the aquifer is $9 \times 10^{-3} \text{m}^2/\text{s}$ and its storage coefficient is 0.05.

Assuming no inflow to the aquifer, use equation 6.4(a) or 6.4(b) to plot the drawdown against r^2/t. Compare the curve obtained with equation 6.10. Calculate the drawdown for $r = 10$ m, 50 m at $t = 1$ hour and $t = 5$ days: also, calculate the radius at which the drawdown is 1 mm after 5 days of pumping.

The variation of piezometric head is given by equation 6.4(b) as

$$\Delta h = \frac{Q}{2\pi T} \int_{\frac{r}{\sqrt{4t(T/S)}}}^{\infty} \frac{1}{u} e^{-u^2}\, du$$

in which

$$\frac{Q}{2\pi T} = \frac{7 \times 10^{-3}}{2 \times \pi \times 9 \times 10^{-3}} = 0.1238 \text{ m}$$

and

$$\frac{S}{4T}\frac{r^2}{t} = \frac{0.05}{4 \times 9 \times 10^{-3}}\frac{r^2}{t} = 1.3889\frac{r^2}{t}$$

Values of the integral in the following table have been obtained by linear interpolation between successive values from the table of values of the integral on pp. 194–203.

r^2/t	$\sqrt{1.3889\dfrac{r^2}{t}}$	$\int_{\sqrt{1.3889\, r^2/t}}^{\infty} \frac{1}{u} e^{-u^2}\, du$	Δh
10^{-4}	0.011785	4.153015	0.514
10^{-3}	0.037268	3.001773	0.372
10^{-2}	0.117851	1.856650	0.230
10^{-1}	0.372678	0.765540	0.0948
1	1.178511	0.059098	0.00732

Using equation 6.10

$$\Delta h = 2.3\frac{Q}{4\pi T}\log\frac{2.25Tt}{r^2 S}$$

$$= \frac{Q}{2\pi T}\ln\sqrt{\frac{2.25Tt}{r^2 S}}$$

$$\frac{2.25T}{S} = \frac{2.25 \times 9 \times 10^{-3}}{0.05} = 0.405$$

$$\Delta h = 0.1238 \ln\sqrt{\frac{0.405}{r^2 t}}$$

r^2/t	10^{-4}	10^{-3}	10^{-2}	10^{-1}	1
Δh	0.514	0.372	0.229	0.0866	–0.0559

The two functions are plotted on Fig. 6.3.

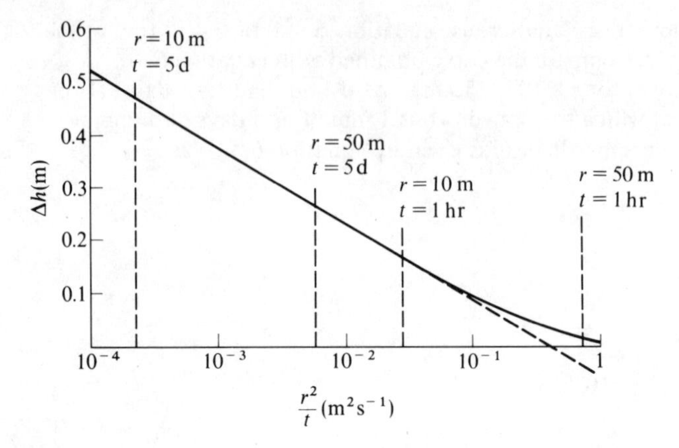

Fig. 6.3. Drawdown in a confined aquifer.

For $r = 10$ m and 50 m at $t = 1$ hour and 5 days, values of r^2/t are as follows:

t \ r	10 m	50 m
1 hr ($= 3.6 \times 10^3 s$)	0.027778(a)	0.694444(b)
5 d ($= 4.32 \times 10^5 s$)	0.0002315(a)	0.005787(a)

Corresponding values of Δh are shown in the following table:

t \ r	10 m	50 m
1 hr	0.166 m	0.0360 m
5 d	0.462 m	0.263 m

where values (a) are used in equation 6.10 and (b) in equation 6.4(b) as indicated by Fig. 6.3.

For $\Delta h = 1$ mm $= 10^{-3}$ m, Fig. 6.3 shows that equation 6.4 is required:

$$\frac{\Delta h}{Q/2\pi T} = \int_{\sqrt{1.3889\, r^2/t}}^{\infty} \frac{1}{u} e^{-u^2}\, du = \frac{10^{-3}}{0.1238} = 0.0080775$$

From the table on pp. 194–203 this value of the integral lies between those for values of the argument of 1.684 and 1.683. By linear interpolation

$$\sqrt{1.3889\, r^2/t} = 1.6831$$

$$\frac{r^2}{t} = 2.0396 \text{ m}^2/\text{s}$$

Since $t = 4.32 \times 10^5$ s

$$r = 938.67 \text{ m}$$

7
Flow in leaky aquifers

Many natural aquifers are neither confined nor unconfined but leaky. By this it is meant that the confining layers are not entirely impervious. For example, one aquifer may be overlain by another and separated from it by a stratum whose permeability is much less than that of either aquifer but is not zero. Analysis of this problem was undertaken first in Holland and later in the United States by Jacob (1946). Probably the most significant single contributor is Hantush, some of whose papers are referred to here (Hantush (1956, 1959b, 1960, 1962b, 1964)). However, many other authors from several countries have worked in this field.

Leakage is a distributed flow to or from the aquifer under study and this has to be allowed for in the equation of continuity. With reference to Fig. 7.1 consideration of the flow into and from a prismatic element of the lower aquifer leads to the equation

$$-\left(\frac{\partial u}{\partial x}\Delta x\right)b\Delta y - \left(\frac{\partial v}{\partial y}\Delta y\right)b\Delta x + K_1\frac{H-h}{b_1}\Delta x\,\Delta y = 0$$

The first two terms are the net rates of flow from the elements parallel to OX and OY and the third term is the leakage flow through its top end. The leakage is into the lower aquifer for $H > h$ and from it for $H < h$.

It is assumed that u and v do not vary with z and that w is negligible (c.f. Chapter 2). Also, flow through the separating stratum is assumed to be vertical – that is refraction through 90° at the interface is implied which is reasonable if the ratio $K : K_1$ is large (section 5.9).

Hence

$$-b\frac{\partial u}{\partial x} - b\frac{\partial v}{\partial y} + \frac{K_1}{b_1}(H-h) = 0$$

and using Darcy's law, we obtain

$$bK\nabla^2 h - \frac{K_1}{b_1}(h-H) = 0$$

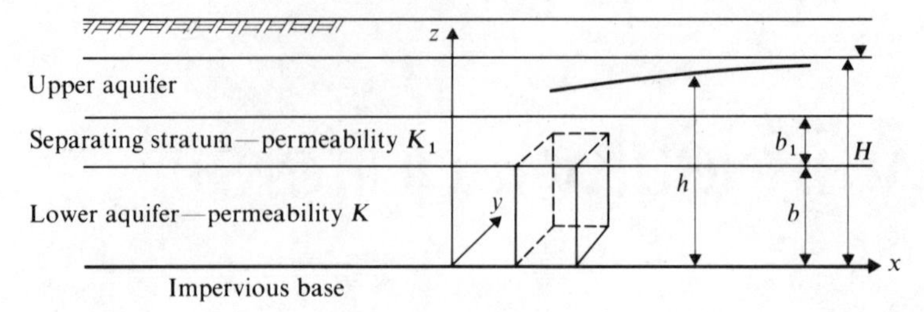

H is piezometric head in upper aquifer
h is piezometric head in lower aquifer

Fig. 7.1. Diagram of a leaky aquifer.

or

$$\nabla^2 h - \frac{K_1}{b_1 bK}(h - H) = 0 \qquad 7.1$$

Equation 7.1 is readily modified to describe unsteady flow, as follows:

$$\nabla^2 h - \frac{K_1}{Kbb_1}(h - H) = \frac{S}{T}\frac{\partial h}{\partial t} \qquad 7.2$$

This complete equation contains one term, $\nabla^2 h$, describing steady flow without leakage, a linear leakage term and a term describing changes in storage in the lower aquifer.

Solution is facilitated by using the notation of Hantush (1959b). The drawdown s is defined by

$$s = H - h$$

and the leakage factor B by

$$B = \sqrt{Kbb_1/K_1}$$

Equation 7.2 becomes

$$\nabla^2 s - \frac{s}{B^2} = \frac{S}{T}\frac{\partial s}{\partial t} \qquad 7.3$$

With various boundary conditions, this equation can be used to describe some flows of practical interest. The value of the solutions depends on the validity of the assumptions used, but generally answers that are at least useful approximations can be obtained.

We consider first a steady two-dimensional flow. Fig. 7.2(*a*) is a cross section of a river in flood and confined to its channel by a levee or stopbank. Water flows from the channel in the permeable stratum and leaks upwards through the overburden and the pastures above are waterlogged to ground level. Fig. 7.2(*b*) shows an approximately equivalent flow for which equation 7.3 would yield

$$\frac{\partial^2 s}{\partial x^2} - \frac{s}{B^2} = 0 \qquad 7.4$$

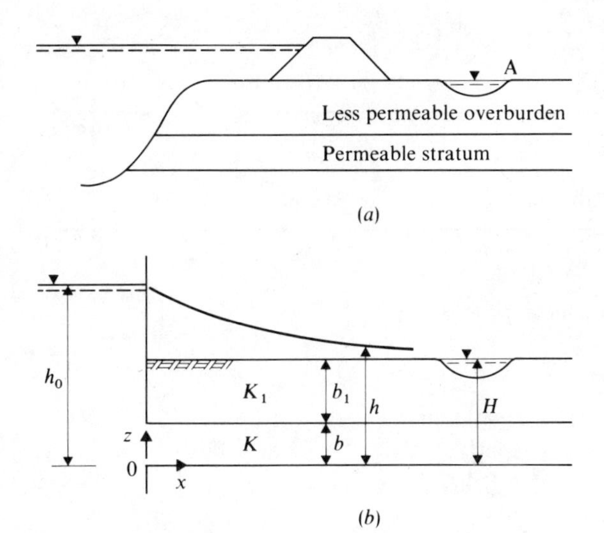

Fig. 7.2. Application of leaky-aquifer theory to a stopbank problem.

The boundary conditions are

$$x = 0, \quad s = H - h_0$$

$$x = \infty, \quad s = 0$$

and the solution is

$$h - H = (h_0 - H)\, e^{-(x/B)} \qquad 7.5$$

The most interesting practical feature of this flow is its effect on places where the weight of overburden is reduced locally by a depression in the surface, as at A in Fig 7.2(*a*). At such a place the exit pressure gradient (ref. section 4.2) may exceed the floatation gradient and be sufficient to make an opening by lifting the overburden. This would allow the flood water to escape from the river. Such an overburden failure caused a major flood in the Bay of Plenty, New Zealand, while the stopbanks remained intact.

Example 7.1

Fig. 7.3 shows a stopbank designed to confine a flooded river to its channel. A stratum of sand 1.5 m thick lies under one of tight silt as shown, and 450 m from the river there is a depression 0.6 m deep in the surface of the land, which is otherwise level. Dimensions and hydraulic properties of the strata are given on Fig. 7.3. The porosity of the silt is 55% and the specific gravity of its solids is 2.65. Examine the possibility of a blow out through the depression.

$$B = \sqrt{\frac{Kbb_1}{K_1}} = \sqrt{\frac{1.5 \times 10^{-2} \times 1.5 \times 1.0}{9 \times 10^{-8}}} = 500 \text{ m}$$

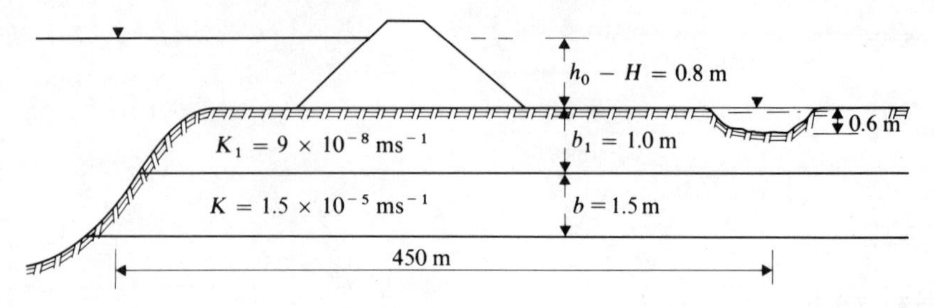

Fig. 7.3. Stopbank on layered soil.

From equation 7.5

$h - H = (h_0 - H)\, e^{-x/B}$; $\quad h_0 - H = 0.8$ m

$\dfrac{x}{B} = \dfrac{450}{500} = 0.9$; $\qquad e^{-x/B} = 0.407$

Hence, at $x = 450$ m, $h - H = 0.8 \times 0.407 = 0.326$ m

$$\text{Exit gradient to depression} = \frac{0.326}{1.0 - 0.6} = 0.815$$

From section 4.2 floatation gradient is

$i_f = (1 - n)(S_s - 1) = 0.45 \times 1.65 = 0.72 < 0.815$

i.e. a blow out is to be expected.

Another two-dimensional flow of interest is the flow to a gallery (a slot in the aquifer, instead of circular hole) at $x = 0$ in the lower aquifer. The upper water table is horizontal and equation 7.4 is again relevant. The boundary conditions are

$x = 0$, $s = s_0 = H - h_0$

$x = \infty$, $s = 0$

where H is the elevation of the upper water table and s_0 and h_0 are the drawdown and piezometric head at the gallery. Equation 7.5 is the solution to this problem also, although more convenient with signs changed on both sides:

$$H - h = s = (H - h_0)\, e^{-x/B}$$

If q_w is the flow rate per unit length into the gallery from both sides and q is the flow rate in the aquifer at distance x from the gallery

$$q = -Kb\frac{\mathrm{d}h}{\mathrm{d}x} = Kb\frac{\mathrm{d}s}{\mathrm{d}x} = K\frac{b}{B}(H - h_0)\, e^{-x/B}$$

and since $q = q_w/2$ at $x = 0$

$$q = \frac{q_w}{2}\, e^{-x/B}$$

When the upper water table is not horizontal – if for example, $H = a + mx$ – equation 7.1 is satisfied by

$$h = C_1 e^{-x/B} + C_2 e^{x/B} + a + mx$$

and C_1 and C_2 are determined by the boundary conditions.

Multilayered aquifers can be treated similarly. For example, the top layer might be a free surface water table, a second layer might be used as a source of supply with a third and lower layer separated from it by a leaky stratum. Flow in each of these layers affects the flow in the other and simultaneous solution of equations for the flow rates is required.

If a well is drilled through to the impervious base in Fig. 7.1, the upper lengths being cased, water can be pumped from the lower aquifer. This flow, which has radial symmetry, has been analysed by Hantush and Jacob (1955). Their solution is given in several forms as follows:

$$s = \frac{Q}{4\pi T} \int_m^\infty \frac{1}{x} e^{-x - \frac{r^2}{4B^2 x}} \mathrm{d}x \qquad 7.6(a)$$

$$= \frac{Q}{4\pi T} \left[2 K_0 \left(\frac{r}{B} \right) - \int_n^\infty \frac{1}{x} e^{-x - \frac{r^2}{4B^2 x}} \mathrm{d}x \right] \qquad 7.6(b)$$

$$= \frac{Q}{4\pi T} I \left(m, \frac{r}{B} \right) \qquad 7.6(c)$$

In equation 7.6(a) the dummy variable x is not a space coordinate – it arises from several transformations in the analysis. The lower limit m is, as before, the group $(r^2 S)(4Tt)$. Equation 7.6(b) is useful because it separates the steady flow which is achieved as a limit at $t = \infty$. In this equation, K_0 is the modified Bessel function of the second kind and of zero order. The lower limit n is given by

$$n = \frac{r^2}{4B^2} \frac{1}{m} = \frac{Tt}{SB^2}$$

Clearly, as $t \to \infty$ the second term on the right in equation 7.6(b) tends to zero and the ultimate steady flow is given by

$$s = \frac{Q}{2\pi T} K_0 \left(\frac{r}{B} \right) \qquad 7.7$$

The integral in equation 7.6(a) is a function of r/B and m and can be evaluated once and for all, the results being presented in tables and graphs (de Wiest (1965)). This gives rise to the form of solution in equation 7.6(c) for which values of $I(m, r/B)$ are found in the tables.

The formation constants may be found by analysing the results of a pumping test. As in the case of unsteady flow without leakage, a graphical superposition is possible, the relevant equations being

$$s = \frac{Q}{4\pi T} I \left(m, \frac{r}{B} \right)$$

whence

$$\log s = \log \frac{Q}{4\pi T} + \log \left[I \left(m, \frac{r}{B} \right) \right]$$

and

$$m = \frac{r^2 S}{4Tt}$$

whence

$$\log t = \log \frac{r^2 S}{4T} + \log \frac{1}{m}$$

for a particular value of r. Observed values of s are plotted against t on log-log scales for this observation well. The function I can be plotted against $1/m$, also on log-log scales, in the form of a family of curves for a range of values of r/B. When the two are superimposed, the shift of origin gives $Q/4\pi T$ and $r^2 S/4T$, (c.f. Fig. 6.1). The curve from the family I versus m and r/B which fits the observations gives r/B. Thus, T, S and B may be calculated (Cooper and Jacob (1946), Hantush (1959b), Engelund (1970)).

Hantush and Jacob (1954) applied the superposition technique to steady flow using equation 7.7. Taking logarithms of both sides yields

$$\log s = \log \frac{Q}{2\pi T} + \log K_0 \left(\frac{r}{B} \right)$$

We also have

$$\log r = \log B + \log \frac{r}{B}$$

When K_0 is plotted against r/B and s against r, using log-log scales of the same size, the shift of the origin gives $Q/2\pi T$ and B. Note that S cannot be found, the flow being steady.

An alternative analysis for steady flow is based on an approximation for equation 7.7. If $r/B < 0.05$ that equation is approximately the same as

$$s = \frac{2.3Q}{2\pi T} \log \frac{0.89r}{B}$$

If s is the plotted against $\log r$, a straight line can be drawn to fit the points for small enough values of r. The slope of this line will give $2.3Q/2\pi T$ and hence T. If the line is extrapolated to $s = 0$, the intercept r_0 will give B since

$$\frac{0.89r_0}{B} = 1$$

Hantush (1956) has also presented a method of analysis of unsteady flow data, a technique which does not require graphical superposition. Observed values of s at a particular value of r are plotted against $\log t$. The curve is

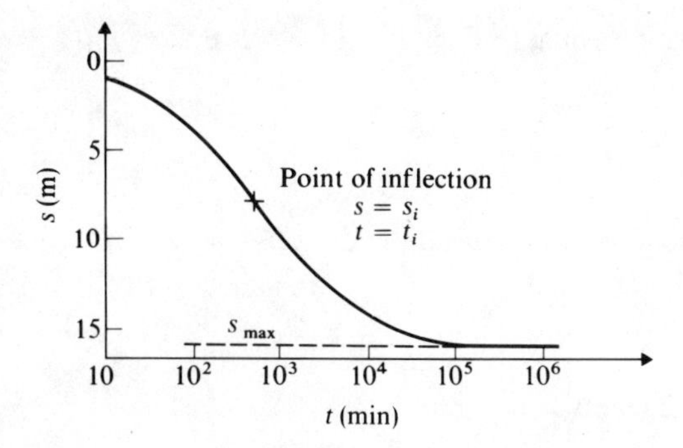

Fig. 7.4. Drawdown in a leaky aquifer as a function of time.

S-shaped (Fig. 7.4) and the asymptote $s = s_{max}$ can usually be plotted quite accurately. Hantush studied the coordinates of the inflection point and the slope of the curve there.

From equation 7.6(a) the slope at any point is

$$a = \frac{ds}{d(\log t)} = 2.3\frac{Q}{4\pi T} e^{-m - \frac{r^2}{4B^2 m}}$$

and

$$\frac{d^2 s}{d(\log t)^2} = \frac{2.3^2 Qm}{4\pi T}\left(1 - \frac{r^2}{4B^2 m^2}\right) e^{-m - \frac{r^2}{4B^2 m}}$$

The point of inflection is given by

$$m = \frac{r}{2B} \qquad 7.8$$

The point of inflection is at time t_i, so that

$$\frac{r^2 S}{4Tt_i} = \frac{r}{2B}$$

and

$$t_i = \frac{BrS}{2T} \qquad 7.9$$

The other coordinate of the point of inflection, s_i, is found by substituting from equation 7.8 in equation 7.6(a)

$$s_i = \frac{Q}{4\pi T}\int_{\frac{r}{2B}}^{\infty} e^{-x - \frac{1}{x}\left(\frac{r}{2B}\right)^2}\, dx$$

and it can be shown that the integral is equal to $K_0(r/B)$. It follows that

$$s_i = \frac{Q}{4\pi T} K_0 \left(\frac{r}{B}\right) = \frac{1}{2} s_{max} \qquad 7.10$$

using equation 7.7

The slope of the curve at the point of inflection, a_i, is given by

$$a_i = \frac{2.3Q}{4\pi T} e^{-(r/B)} \qquad 7.11$$

and this divided into equation 7.10 yields

$$\frac{2.3s_i}{a_i} = K_0 \left(\frac{r}{B}\right) e^{r/B} \qquad 7.12$$

These relationships are applied to the problem by locating the point of inflection (Fig. 7.4) using equation 7.10. The slope of the curve at that point is easily found. Hence, using these values of s_i and a_i, the left side of equation 7.12 may be calculated. The right side of equation 7.12 is a function of r/B only and tables of the function are available (Hantush (1956), De Wiest (1965)). Thus r/B and B can be found. With r/B and a_i known, T follows from equation 7.11 and S may be calculated using equation 7.9.

If the constants so calculated are used in equation 7.6 to calculate drawdown as a function of time, the computed curve usually fails to coincide with the observed line in the early stages of pumping. Hantush discusses this point and such special problems of high rates of leakage and extrapolation to s_{max}.

Example 7.2

Water is pumped from a well in a leaky aquifer at the rate of 95 l/s. The drawdown of the piezometric surface in the aquifer is measured in an observation well 325m from the well being pumped, the results being as follows:

$t(s)$	1.45×10^3	2.9×10^3	7.25×10^3	1.45×10^4
$s(m)$	1.65	2.22	3.02	3.62
t	2.9×10^4	7.25×10^4	1.45×10^5	2.9×10^5
s	4.20	4.94	5.40	5.76
t	7.25×10^5	1.45×10^6	2.9×10^6	
s	6.00	6.05	6.05	

Determine the following parameters for the leaky aquifer:

(1) the transmissivity $T(=Kb)$

(2) the length $B(=\sqrt{Kbb'/K'})$ and hence K'/b'

(3) the storage coefficient S.

A brief table of $e^x K_0(x)$ is as follows:

x	= 0.035	0.036	0.037	0.038	0.039
$e^x K_0(x)$	= 3.5933	3.5678	3.5430	3.5189	3.4955
x	= 0.040	0.041	0.042	0.043	0.044
$e^x K_0(x)$	= 3.4727	3.4505	3.4289	3.4079	3.3874
x	= 0.045	0.046	0.047	0.048	0.049
$e^x K_0(x)$	= 3.3673	3.3478	3.3287	3.3100	3.2918

where $x = r/B$.

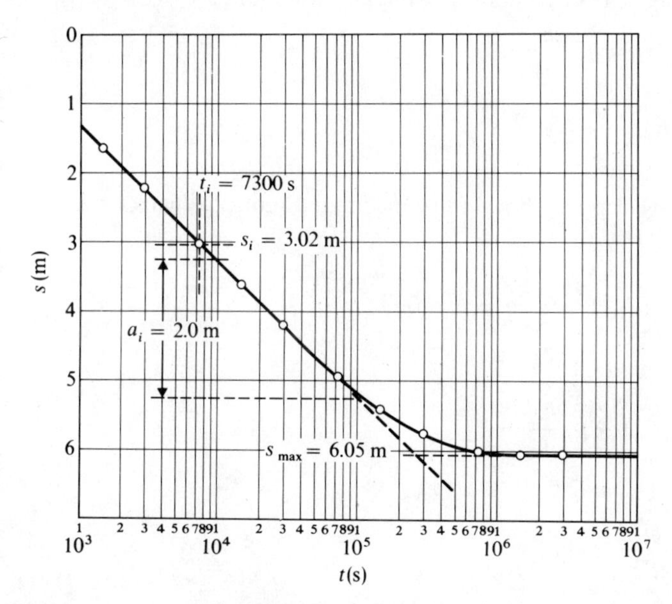

Fig. 7.5. Determination of formation constants from pump test data of a well in a leaky aquifer.

From Fig. 7.5 $s_{max} = 6.05$ m, and over one log-cycle the slope of the straight line is $a_i = 2.0$ m. From equation 7.10 $s_i = 3.02$ m and inserting in equation 7.12

$$\frac{2.3 s_i}{a_i} = 3.47 = e^{r/B} K_0 \left(\frac{r}{B} \right)$$

Interpolating from the table yields $r/B = 0.0401$ or $B = 325/0.0401 = 8100$ m.

(1) From equation 7.11

$$T = \frac{2.3Q}{4\pi a_i e^{r/B}} = \frac{2.3 \times 9.5 \times 10^{-2}}{4\pi \times 2 \times 1.042} = 8.33 \times 10^{-3} \text{ m}^2/\text{s}$$

(2) From

$$B = \sqrt{Kbb'/K'} = \sqrt{T(K'/b')}$$

$$\frac{K'}{b'} = \frac{T}{B^2} = \frac{8.33 \times 10^{-3}}{8.1^2 \times 10^6} = 1.27 \times 10^{-10} s^{-1}$$

(3) From equation 7.9

$$S = \frac{2Tt_i}{Br} = \frac{2 \times 8.33 \times 10^{-3} \times 7.3 \times 10^3}{8.1 \times 10^3 \times 3.25 \times 10^2} = 4.61 \times 10^{-5}$$

Flow in a leaky aquifer may also be associated with a gravity flow, as shown in Fig. 7.6, for example. In this case water is being pumped from a well in an upper unconfined aquifer and there is recharging from above by rainfall and from below by flow through a leaky stratum. The relevant equations are

$$Q = 2\pi rKh_1 \frac{dh_1}{dr} \simeq 2\pi rK\bar{h}_1 \frac{dh_1}{dr}$$

$$dQ = -2\pi rdr \left[P + \frac{(h_2 - h_1)K'_1}{b'_1}\right]$$

where $\bar{h}_1$ is the average depth of flow in the unconfined aquifer. When $Q = 0$, $h_2 = h_0$. Hence,

$$\frac{d^2h_1}{dr^2} + \frac{1}{r}\frac{dh_1}{dr} - \frac{K'_1}{K\bar{h}_1 b'_1} h_1 + \frac{Pb'_1 + K'_1 h_0}{K\bar{h}\, b'_1} = 0$$

and the general solution is

$$h_1 = C_1 I_0\left(\frac{r}{B}\right) + C_2 K_0\left(\frac{r}{B}\right) + \frac{Pb'_1}{K'_1} + h_0$$

Since $h_1 \to h_0$ as $r \to \infty$ and $Q = Q_w$ at $r = r_w$

$$C_1 = 0, \quad C_2 = -\frac{Q}{2\pi K\bar{h}_1}$$ and I_0 is

the modified Bessel function of zero order. Thus,

$$h_1 = \frac{Pb'_1}{K'_1} + h_0 - \frac{Q}{2\pi K\bar{h}_1} K_0\left(\frac{r}{B}\right)$$

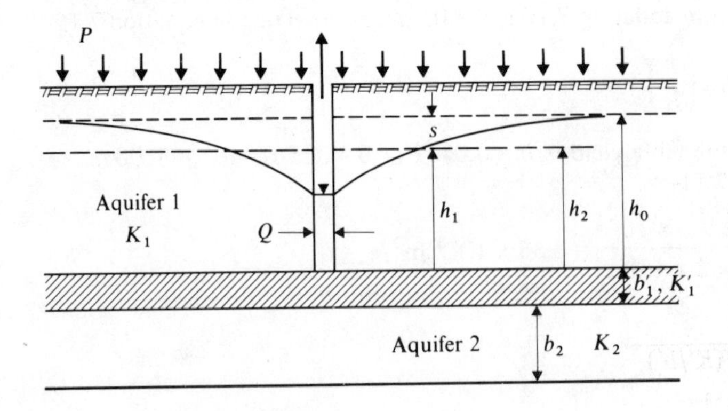

Fig. 7.6. Gravity flow and leaky aquifer.

The drawdown is s, given by

$$s = \frac{Pb'_1}{K'_1} + h_0 - h_1$$

whence

$$s = \frac{Q}{2\pi K\bar{h}_1} K_0\left(\frac{r}{B}\right) \simeq \frac{Q}{2\pi K\bar{h}_1} \ln \frac{1.123B}{r}$$

if r/B is small.

This analysis can be extended to describe flow in the same series of strata (Fig. 7.6) but with a unidirectional flow in each aquifer as well. If the well penetrates both aquifers and the approximation of constant flow depth is used again for the unconfined aquifer then

$$Q_1 = 2\pi r K_1 \bar{h}_1 \frac{dh_1}{dr}$$

$$Q_2 = 2\pi r K_2 b_2 \frac{dh_2}{dr}$$

$$dQ_1 = -2\pi r dr \left(P - \frac{h_1 - h_2}{b'_1/K'_1}\right)$$

and

$$dQ_2 = -2\pi r dr \frac{h_1 - h_2}{b'_1/K'_1}$$

Hence,

$$\frac{d^2(h_1 - h_2)}{dr^2} + \frac{1}{r}\frac{d(h_1 - h_2)}{dr} - \frac{h_1 - h_2}{B^2} = -\frac{P}{K_1\bar{h}_1}$$

$$\frac{d^2(\alpha h_1 + h_2)}{dr^2} + \frac{1}{r}\frac{d(\alpha h_1 + h_2)}{dr} = -\frac{\alpha P}{K_1\bar{h}_1}$$

where

$$\alpha = K_1\bar{h}_1/K_2 b_2, \quad \frac{1}{B^2} = \frac{K'_1}{K_1\bar{h}_1 b'_1} + \frac{K'_1}{K_2 b_2 b'_1}$$

These equations are satisfied by

$$h_1 - h_2 = C_1 I_0\left(\frac{r}{B}\right) + C_2 K_0\left(\frac{r}{B}\right) + \frac{PB^2}{K_1\bar{h}_1}$$

$$\alpha h_1 + h_2 = C_3 \ln r - \frac{\alpha P}{4K_1\bar{h}_1} r^2 + C_4$$

and the constants C_1 to C_4 are determined by the boundary conditions.

Analysis of flow in a leaky aquifer of infinite extent can be adapted to a semi-infinite aquifer with a boundary at $x = 0$ by applying the method of images.

For the infinite aquifer we have

$$s = \frac{Q}{2\pi Kb} K_0\left(\frac{r}{B}\right)$$

Superposition of the drawdown due to an image well yields for the combination

$$s_s = \frac{Q}{2\pi Kb}\left[K_0\left(\frac{r}{B}\right) - K_0\left(\frac{r_i}{B}\right)\right]$$

At the well face $r = r_w$ and $r_i = 2x_1$, the well being at $x = x_1$ and its image at $x = -x_1$. Hence, the drawdown at the well is

$$s_{s0} = \frac{Q}{2\pi Kb}\left[K_0\left(\frac{r_w}{B}\right) - K_0\left(\frac{2x_1}{B}\right)\right]$$

A three-layered problem can be posed, in which another leaky aquifer underlies the one from which water is pumped. Darcy's law and the continuity requirement lead to

$$\frac{d^2 s_2}{dr^2} + \frac{1}{r}\frac{ds_2}{dr} - \frac{K_1' s_2}{K_2 b_2 b_1'} - \frac{K_2'(s_2 - s_3)}{K_2 b_2 b_2'} = 0$$

$$\frac{d^2 s_3}{dr^2} + \frac{1}{r}\frac{ds_3}{dr} - \frac{K_2'(s_3 - s_2)}{K_3 b_3 b_2'} = 0$$

where s_2 is the drawdown in the middle layer and K_1', b_1', K_2', b_2' are permeabilities and thicknesses for the upper and lower leaky layers.

The drawdown s_2 is given by

$$s_2 = \frac{Q}{2\pi K_2 b_2}\frac{1}{B_2 - B_3}\left[(B_2 - \alpha_3)K_0(\sqrt{B_2}r) + (\alpha_3 - B_3)K_0(\sqrt{B_2}r)\right]$$

where $\alpha_2 = K_1'/K_2 b_2 b_1'$, $\alpha_3 = K_2'/K_2 b_2 b_2'$. When $\sqrt{B_2}r$ is small, the Bessel function $K_0(\sqrt{B}r) \simeq \ln(1.123/\sqrt{B}r)$; Huisman and Kemperman (1951).

Engelund (1970) has suggested separating the process of pumping from a leaky aquifer into three time periods. In the first, immediately after the start of pumping the flow of groundwater into the well is small and the drawdown is rapid. During the second period pore water pressure in the soil falls and water stored in the aquifer by compression flows to the well. A third period may occur in which leakage from adjacent strata is dominant.

8 Unsteady flow in unconfined aquifers

Unsteady flow in unconfined aquifers

This unsteady flow is more complicated than those examined in Chapter 6. In the first place the water table, which is the upper boundary of the effective aquifer, varies with time. Associated with this are vertical components of velocity which cannot be ignored, particularly near the well. The boundary condition at the well also raises mathematical difficulties. In the early stages of pumping, the water level in the well is lower than the water table outside the well and there is a surface of seepage on the well face. The extent of this surface varies with time as the water table varies and, if the pumping rate is constant, as the water level in the well changes. If the water level in the well is kept constant the surface of seepage still changes with time and the flow rate varies. Significant aspects of the flow are sketched on Fig. 8.1. The water table is not a stream surface in this flow and the surface of seepage is neither a streamline nor an equipotential. Below the surface of the water in the well, the well face is an equipotential and all stream surfaces meeting it do so at right angles.

The volume of the cone of depression at any time is related to the total volume of water pumped. From the derivation of equation 6.2 it is clear that the total volume pumped from a confined aquifer equals the volume of the cone of depression in the piezometric surface multiplied by the storage coefficient S. In that case, the water is supplied by the changes in deformation of the water and the aquifer. In the present case, corresponding changes in strain occur and the quantity of water yielded in this fashion is the same. However, an additional larger quantity is released by drainage of the voids in the cone of depression. These voids will not be completely drained and the volume of water derived from the cone of depression in this way will be less than the total volume of the voids. Some water is retained in the voids as described in Chapter 1. As this water is released it flows through the water table to rejoin the main flow. Boulton (1963) has referred to this phenomenon as delayed yield and it also has aspects resembling flow to a leaky aquifer.

For unconfined flow, the storage coefficient S is given by

$$S = S_v + S_e$$

where S_v = storage coefficient due to partial drainage of voids; it will approach the porosity of the aquifer as a limit

S_e = elastic storage coefficient

$= h_e \rho g(\alpha + n\beta)$

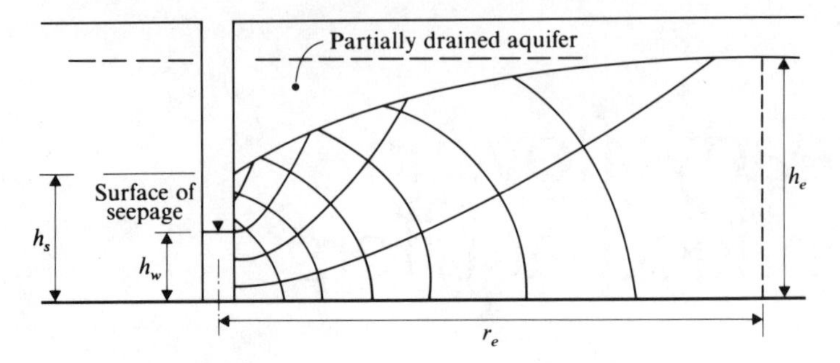

Fig. 8.1. Diagrammatic flow net for unsteady flow in an unconfined aquifer.

from equation 2.18. The ratio of S_e to S_v can be shown to be small by estimating S_e:n if S_v and n are the same order of size. Thus,

$$\frac{S_e}{n} = h_e \rho g \left(\beta + \frac{\alpha}{n} \right)$$

Both α and β are much smaller than $h_e \rho g$ unless h_e is extremely large. Consequently, it is a valid approximation to ignore changes in the deformation of the water and the aquifer. The storage coefficient is then determined by the yield from drainage of voids.

The equation to be solved for the water filled part of the aquifer is the Laplace equation derived from equation 6.1 *viz.*

$$\nabla^2 h = \frac{S_e}{T} \frac{\partial h}{\partial t} = 0$$

i.e. $$\frac{\partial^2 h}{\partial r^2} + \frac{1}{r} \frac{\partial h}{\partial r} + \frac{\partial^2 h}{\partial_z{}^2} = 0 \qquad 8.1$$

The boundary conditions to be satisfied by the solution are as follows (Boulton (1954, 1965)):

(1) At the water table

$$p = 0$$

so that $h - z = 0$

Since the normal components of the velocity of the water table and an element of water in the water table are equal (Raudkivi and Callander (1975)),

$$\left(\frac{\partial}{\partial t} + \frac{v_r}{S} \frac{\partial}{\partial r} + \frac{v_z}{S} \frac{\partial}{\partial z} \right) (h - z) = 0$$

Hence, at the water table

$$\frac{\partial h}{\partial t} = \frac{K}{S} \left\{ \left(\frac{\partial h}{\partial r} \right)^2 + \left(\frac{\partial h}{\partial z} \right)^2 - \frac{\partial h}{\partial z} \right\}$$

The squares of the gradients of h can be discarded to yield the linearized boundary condition

$$z = h$$

$$\frac{\partial h}{\partial t} + \frac{K}{S}\frac{\partial h}{\partial z} = 0 \qquad 8.2$$

(2) At the lower boundary where there is a horizontal impermeable surface

$$z = 0$$

$$\frac{\partial h}{\partial z} = 0 \qquad 8.3$$

(3) At the well face, the hydrostatic distribution of pressure below water level in the well gives

$$r = r_w, \quad 0 \leqq z \leqq h_w,$$

$$h = h_w \qquad 8.4$$

For the surface of seepage the pressure is zero and

$$r = r_w, \quad h_w \leqq z \leqq h_s,$$

$$h = z \qquad 8.5$$

The problem has been solved by Boulton for Q = constant (Boulton (1954) for h_w = constant (Boulton (1965)).

In the first paper he replaced the boundary conditions equations 8.2, 8.4 and 8.5 by approximations. Instead of satisfying equation 8.2 at $z = h$ he satisfied it at $z = h_e$. That is, the solution satisfies the boundary condition at the initial water table instead of the time varying surface. He also replaced the boundary conditions at the surface of the well by assuming that the constant flow rate Q is uniformly distributed along a length h_e of a well of infinitesimal radius. That is

$$r \to 0, \quad 0 \leqq z \leqq h_e,$$

$$Q = 2\pi K h_e r \frac{\partial h}{\partial r} \qquad 8.6$$

The solution he obtained is

$$h_e - h = \frac{Q}{2\pi K h_e} \int_0^\infty \frac{J_0\left(u\frac{r}{h_e}\right)}{u} \left\{1 - \frac{\cosh\left(u\frac{z}{h_e}\right)}{\cosh u} \exp(-\tau u \tanh u)\right\} du \qquad 8.7$$

where

$$\tau = \frac{Kt}{Sh_e}$$

u is a dummy variable which disappears on integration between limits and J_0 is the Bessel function of the first kind of zero order.

Equation 8.7 gives the distribution of piezometric head h as a function of r, z and t.

A first approximation to the drawdown Δh is obtained by substituting $z = h_e$

and Boulton showed that

$$\Delta h = \frac{Q}{4\pi T}\{-Ei(-m) - 2X_1\}$$

where Ei is the exponential integral defined in Chapter 6.

$$T = Kh_e$$

$$m = \frac{r^2 S}{4Tt} \qquad 6.5$$

and X_1 is a correction term which is a function of the dimensionless variables r/h_e and τ. Boulton deduced that X_1 may be neglected for

$$\tau > 5 \qquad 8.8$$

so that, after a long enough time, the drawdown in this flow is given by

$$\Delta h = \frac{Q}{4\pi T}\{-Ei(-m)\}$$

i.e.

$$\Delta h = \frac{Q}{4\pi T} I \qquad 6.6$$

Thus, subject to the inequality 8.8. the same equation gives the drawdown for unconfined flow as well as confined.

This might have been anticipated from the simplified boundary conditions. After sufficient time has passed, the surface of seepage will disappear and equation 8.6 corresponds to a well penetrating an aquifer of thickness h_e. This boundary condition then resembles that for confined flow.

As a result, the methods of Chapter 6 can be used to find the formation constants for unconfined flow also.

In the second paper Boulton solved the problem for h_w = constant. For this case the boundary conditions described by equations 8.4 and 8.5 can be satisfied and the only approximation required is to satisfy equation 8.2 at $z = h_e$ instead of $z = h$. The solution for the variable flow rate is

$$Q = \frac{8}{\pi} Th_e \int_0^\infty \frac{1}{u^3} \frac{1}{J_0^2\left(u\frac{r_w}{h_e}\right) + K_0^2\left(u\frac{r_w}{h_e}\right)} \left(1 - \frac{\cosh\left(u\frac{h_w}{h_e}\right)}{\cosh u}\right) \exp(-\tau u \tanh u)\, du \qquad 8.9$$

where J_0 is the Bessel function of the first kind and zero order and K_0 is the Bessel function of the second kind and zero order. The other variables are as defined for equation 8.7.

A dimensionless function G is defined such that

$$Q = \pi Th_e \left[1 - \left(\frac{h_w}{h_e}\right)^2\right] G \qquad 8.10$$

and such that equations 8.9 and 8.10 agree. It is shown that for $\tau > 1$

$$G = \frac{2}{\ln\left(2.25 \dfrac{Tt}{r_w^2 S}\right)} \qquad 8.11$$

When equation 8.11 is used in equation 8.10 the result is

$$Q = \pi K \frac{h_e^2 - h_w^2}{\ln\left(\dfrac{r_e}{r_w}\right)} \qquad 5.19$$

with

$$r_e = 1.5 \sqrt{\frac{Tt}{S}} = 1.5 \sqrt{\frac{Kh_e t}{S}} \qquad 8.12$$

This interesting result shows that the same equation describes both steady and and unsteady flows, the effect of time being completely described by equation 8.12.

Boulton concluded that it would not matter how the drawdown approached this asymptotic form so that equation 5.19 would also be valid for an unsteady flow with Q kept constant, provided $\tau > 1$.

Equations 5.19 and 8.12 offer a means of estimating the formation constants if observations in several wells are available at sufficiently large values of τ. Assuming Q, h_e, h_w, r_w and t are known, there are three variables to be found in the equations: r_e, K and S. If it is assumed that the water table is also given by the steady flow equation we have, from equations 5.19 and 5.20

$$h_e^2 - h^2 = \frac{Q}{\pi K} \ln \frac{r_e}{r}$$

In terms of the drawdown $\Delta h (= h_e - h)$

$$\Delta h \left(1 - \frac{\Delta h}{2h_e}\right) = \frac{Q}{2\pi K h_e} \ln \frac{r_e}{r} \qquad 8.13$$

Thus, if $\Delta h(1 - (\Delta h/2h_e))$ is plotted on a natural scale against r on a logarithmic scale, it should be possible to fit a straight line to the plotted points and to extrapolate it to $\Delta h = 0$ which gives r_e. The equation of the straight line must also agree with equation 8.13 so that $Q/2\pi K h_e$ can be evaluated and this gives K. Equation 8.12 can then be used to find S.

Boulton also examined his solutions for small values of τ and reference should be made to his papers for these results. Using them, pumping tests of short duration can be made in cases where h_e is large or K small. In such circumstances, an unduly large value of t would be required to get τ above 1.

Example 8.1

Results of a pumping test conducted near Kearney, Nebraska in 1933 have been published by Wenzel and Fishel (1942). Water was pumped from a well at 1100

US gal/min for 24 hours. At the end of that time water levels in observation wells were measured and drawdowns computed as follows:

r(ft)	50	100	150	200	250
Δh(ft)	3.93	2.95	2.43	2.05	1.74

The aquifer was not confined above and the initial thickness of the saturated stratum was 48 ft.

Using these data estimate the transmissivity T, the coefficient of permeability K and the storage coefficient S of the aquifer.

We assume that the drawdown is small enough for the water table and the underlying impermeable stratum to be assumed parallel. Provided $\tau > 5$ equations 6.5 and 6.6 can be used. We have

$t = 86\,400$ s

r(ft)	50	100	150	200	250
r(m)	15.25	30.5	45.75	61.0	76.25
r^2/t(m²/s)	2.7×10^{-3}	1.08×10^{-2}	2.42×10^{-2}	4.31×10^{-2}	6.74×10^{-2}
Δh(ft)	3.93	2.95	2.43	2.05	1.74
Δh(m)	1.20	0.900	0.740	0.625	0.530

From tables of $-Ei(-m)$ as a function of m (see, for example, Wenzel and Fishel, 1942):

m	0.001	0.002	0.005	0.01	0.02	0.05	0.1
$I = -Ei(-m)$	6.33	5.64	4.73	4.04	3.35	2.47	1.82

When these two curves are plotted on logarithmic scales and superimposed to fit, the result is as shown on Fig. 8.2. From the measured displacements of the axes,

$$\log \frac{Q}{4\pi T} = -0.690 = \bar{1}.310$$

$$\frac{Q}{4\pi T} = 0.204 \text{ m}$$

$$\log \frac{4T}{S} = +0.200$$

$$\frac{4T}{S} = 1.59 \text{ m}^2/\text{s}$$

From the data, $Q = 1100$ US gal/min $= 69.4$ l/s

Hence,

$$T = \frac{6.94 \times 10^{-2}}{4 \times \pi \times 2.04 \times 10^{-1}} = 0.0270 \text{ m}^2/\text{s}$$

$$S = \frac{4 \times 2.7 \times 10^{-2}}{1.59} = 0.0680$$

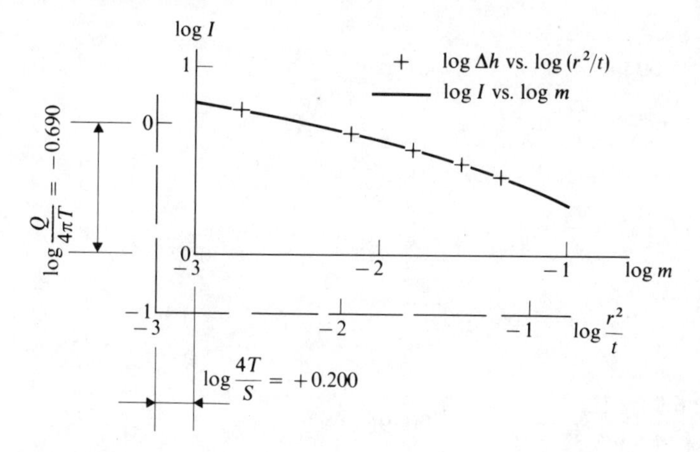

Fig. 8.2. Evaluation of formation constants for unconfined aquifer from pump test data, unsteady flow.

Thickness of aquifer = 48 ft = 14.6 m

$$K = \frac{0.0270}{14.6} = 0.00185 \text{ m/s}$$

$$\tau = \frac{0.00185 \times 86\,400}{0.0680 \times 14.6} = 161 > 5$$

Alternatively, corresponding values of Δh and I, r^2/t and m could be read and used in equations 6.5 and 6.6. In effect, one is then using the coordinates of the common point, referred to the two sets of axes, to find the displacement of the axes.

Using equations 8.12 and 8.13 the same data are treated as follows:

On Fig. 8.3 $\Delta h(1 - h/2h_e)$ is plotted against r on log-normal scales. From this

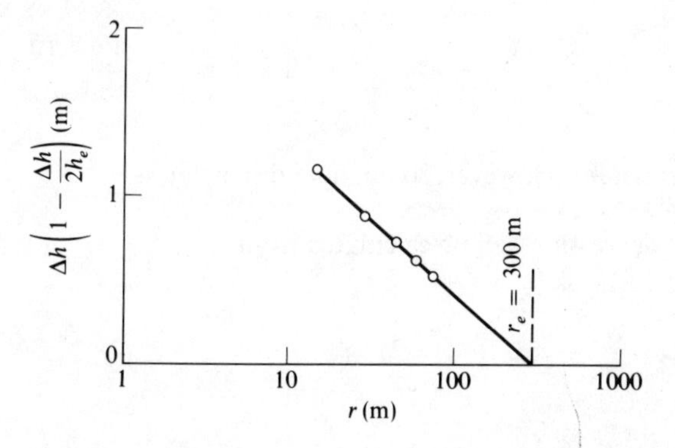

Fig. 8.3. Evaluation of formation constants for unconfined aquifer from pump test data, unsteady flow.

graph

$$r_e = 300 \text{ m}$$

$$\Delta h \left(1 - \frac{\Delta h}{2h_e}\right) = 0.89 \log \frac{r_e}{r} \qquad 8.14$$

i.e.

$$\Delta h \left(1 - \frac{\Delta h}{2h_e}\right) = 0.387 \ln \frac{r_e}{r}$$

Hence, from equation 8.13

$$\frac{Q}{2\pi K h_e} = 0.387$$

$$K = \frac{6.94 \times 10^{-2}}{2 \times \pi \times 14.6 \times 3.87 \times 10^{-1}} = 0.00196 \text{ m/s}$$

From equation 8.12

$$S = \frac{2.25 K h_e t}{r_e^{\,2}}$$

$$= \frac{2.25 \times 0.00196 \times 14.6 \times 86\,400}{300 \times 300}$$

$$= 0.0617$$

Wenzel also published the following observations made in the same test at 16 hours:

r(ft)	50	100	150	200	250
Δh(ft)	3.77	2.78	2.27	1.88	1.59

When these are analysed by the second of the above methods the results are

$$r_e = 250 \text{ m}$$

$$\Delta h \left(1 - \frac{\Delta h}{2h_e}\right) = 0.394 \ln \frac{r_e}{r} \qquad 8.15$$

$$K = 0.00192 \text{ m/s}$$

$$S = 0.0580$$

All three sets of results are consistent. However, some more discussion is justified.

The volume of the cone of depression can be calculated from

$$\mathcal{V} = \int_{r_w}^{r_e} 2\pi r \, \Delta h \, \mathrm{d}r$$

and

$$\Delta h = A \ln \frac{r_e}{r}$$

Hence

$$\mathcal{V} = \frac{1}{2}\pi r_e^2 A$$

provided $(r_w/r_e)^2$ is negligible.

The length A is given approximately by equations 8.14 and 8.15 by noting that the factor $(1 - \Delta h/2h_e)$ does not vary much. For the observations at 16 hours it ranged from 0.961 at $r = 50$ ft. (15.25 m) to 1 at $r = r_e$ and the corresponding range at 24 hours was 0.959 to 1. Taking an average value of 0.98 allows A to be calculated for each set of measurements. We have:

At $t = 16$ hours

$$A = 0.402 \text{ m}$$

$$\mathcal{V} = 4.05 \times 10^4 \text{m}^3$$

$$S\mathcal{V} = 2.28 \times 10^3 \text{ m}^3$$

$$\text{total volume pumped} = 6.94 \times 10^{-2} \text{m}^3/\text{s} \times 57600 \text{ s}$$

$$= 3.99 \times 10^3 \text{m}^3$$

This corresponds to a leakage flow of $1.71 \times 10^3 \text{m}^3$ in 16 hours.

At $t = 24$ hours

$$A = 0.400 \text{ m}$$

$$\mathcal{V} = 5.89 \times 10^4 \text{m}^3$$

$$S\mathcal{V} = 3.69 \times 10^3 \text{m}^3$$

total volume pumped = $6.00 \times 10^3 \text{m}^3$.

The leakage flow in 24 hours amounted to $2.31 \times 10^3 \text{m}^3$.

The results for the storage coefficient are considerably lower than one would expect the porosity to be. The aquifer is described as sand and gravel and although no measurements of porosity were made at Kearney, the porosity of samples from a similarly described aquifer in the same valley at nearby Grand Island was about 30%.

The figures are consistent with the notion that delayed yield or leakage through the water table occurs and they suggest that the leakage flow was nearly steady for the duration of this test. The relative magnitudes of $S\mathcal{V}$ and the delayed yield suggest that delayed yield or leakage should be incorporated in the analysis.

Boulton (1963) did this by considering the effect of lowering the water table by an amount Δs in an interval of time θ to $(\theta + \Delta\theta)$. (In our notation the drawdown $\Sigma\Delta s = \Delta h$.) The amount of water released by this element of change in water table comprises two parts:

(1) Immediate release of a quantity of water $S\Delta s$ per unit of area on plan

(2) Delayed release per unit of area on plan at time $t (t > \theta)$ equal to $\Delta s \alpha S' e^{-\alpha(t - \theta)}$:

The total volume of the delayed yield is $\alpha S' \int_\theta^\infty e^{-\alpha(t-\theta)} dt$ and this equals S'.

Thus, S' is seen to be a storage coefficient for delayed yield; α is a constant and Boulton called its reciprocal the delay index.

The equation solved by Boulton was

$$T\left(\frac{\partial^2}{\partial r^2}+\frac{1}{r}\frac{\partial}{\partial r}\right)\Delta h = S\frac{\partial}{\partial t}\Delta h + \alpha S'\int_0^t \frac{\partial}{\partial t}\Delta h\bigg|_{t=\theta} e^{-\alpha(t-\theta)}d\theta$$

and his solution for small values of t was identical with that obtained by Hantush and Jacob (1955) for a leaky aquifer. We do not pursue the analysis here. The results are of great interest, but there is some doubt about the form assumed for the delayed yield.

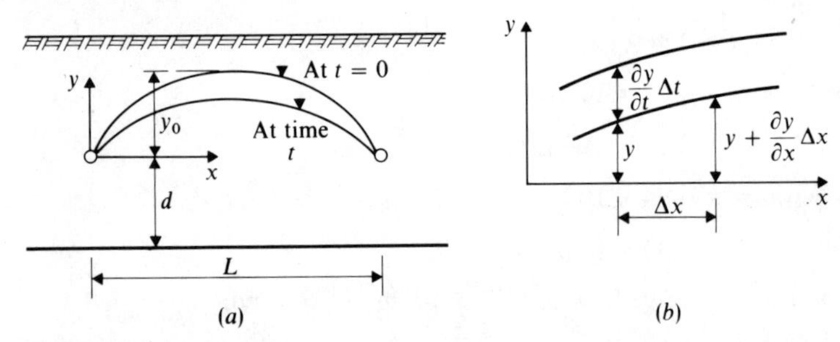

Fig. 8.4. Definition sketch for unsteady unconfined flow to horizontal parallel field tile drains.

Another unsteady unconfined flow of some practical interest is flow to a series of parallel field tiles, as sketched on Fig. 8.4(a). It is assumed that water from a flooded aquifer is being drained by flow to the field tiles. The water table is being lowered as time passes. Darcy's law and the principle of conservation of matter are applied to a control volume of unit width in Fig. 8.4(b) as follows:

$$\text{Volume which flows into control volume} = \left[-K\frac{\partial y}{\partial x}(y+d)\right]\Delta t + \frac{\partial}{\partial x}\left[-K\frac{\partial y}{\partial x}(y+d)\right]\Delta x\Delta t$$

$$\text{Volume which flows from control volume} = \left[-K\frac{\partial y}{\partial x}(y+d)\right]\Delta t$$

$$\text{Decrease in volume stored in control volume} = n\frac{\partial y}{\partial t}\Delta t\Delta x$$

Hence,

$$\frac{\partial}{\partial x}\left[K\frac{\partial y}{\partial x}(y+d)\right] = n\frac{\partial y}{\partial t} \tag{8.16}$$

which is of the form

$$\frac{\partial}{\partial x}\left\{D\frac{\partial y}{\partial x}\right\} = \frac{\partial y}{\partial t},$$ where $D = D(y) = K(y+d)/n$, c.f. equation 1.14.

This non-linear partial differential equation is not readily solved. However, it can be linearized by using a constant average value $\bar{y}$ for the area of flow $y + d$. Then we have

$$K\bar{y}\,\frac{\partial^2 y}{\partial x^2} = n\,\frac{\partial y}{\partial t}$$

or

$$\frac{\partial^2 y}{\partial x^2} = \frac{n}{K\bar{y}}\,\frac{\partial y}{\partial t} \qquad 8.17$$

The solution is obtained by separation of variables. Assuming y to be the product of two functions, one a function of t alone and the other a function of x alone, we have

$$y = \phi_1(t)\phi_2(x)$$

Substituting in equation 8.17

$$\phi_1\phi_2'' = \frac{n}{K\bar{y}}\,\phi_1'\phi_2$$

Hence,

$$\frac{\phi_2''}{\phi_2} = \frac{n}{K\bar{y}}\,\frac{\phi_1'}{\phi_1} = C, \text{ a constant}$$

From

$$\phi_2'' - C\phi_2 = 0$$

$$\phi_2 = A \sin(\sqrt{-C}\,x) + B \cos(\sqrt{-C}\,x)$$

and since

$y = 0$ at $x = 0$ and L

$$B = 0 = A \sin(\sqrt{-C}\,L)$$

Therefore,

$$\sqrt{-C}\,L = n\pi \qquad n = 1, 2, \ldots$$

and taking $n = 1$ as a first approximation

$$\sqrt{-C} = \frac{\pi}{L}$$

Hence,

$$\phi_2 = A \sin\frac{\pi x}{L}$$

From

$$\frac{\phi_1'}{\phi_1} = C\,\frac{K\bar{y}}{n} = -\left(\frac{\pi}{L}\right)^2 \frac{K\bar{y}}{n}$$

$$\ln \phi_1 = -\left(\frac{\pi}{L}\right)^2 \frac{K\bar{y}}{n} t + D$$

$$\phi_1 = \exp\left[-\left(\frac{\pi}{L}\right)^2 \frac{K\bar{y}}{n} t + D\right]$$

Thus,

$$y = \exp\left[-\left(\frac{\pi}{L}\right)^2 \frac{K\bar{y}}{n} t + D\right] \sin\frac{\pi x}{L}$$

If y_0 is the initial water level midway between the drains (Fig. 8.4(*a*)) the constant D can be found by substituting $t = 0, x = L/2$ and $y = y_0$.

We then obtain

$$y = y_0 \exp\left[-\left(\frac{\pi}{L}\right)^2 \frac{K\bar{y}}{n} t\right] \sin\frac{\pi x}{L} \qquad 8.18$$

The highest water level at any time occurs at $x = L/2$ and if this is given by $y = y_m$

$$y_m = y_0 \exp\left[-\left(\frac{\pi}{L}\right)^2 \frac{K\bar{y}}{n} t\right] \qquad 8.19$$

From this, an expression can be found for the spacing L required to achieve a desired drop in water level in a specified time. We have

$$-\left(\frac{\pi}{L}\right)^2 \frac{K\bar{y}}{n} t = \ln\frac{y_m}{y_0}$$

$$L = \pi\sqrt{(K\bar{y}t/n)/\ln(y_0/y_m)} \qquad 8.20$$

The approximate average depth $\bar{y}$ can be assumed to be

$$\bar{y} = d + \frac{1}{2} y_0$$

This is reasonable for relatively large values of d, but even with $d = 0$ agreement with more exact solutions is reasonable for $y_0/y_m \not> 2$.

Another solution, using the same general approach is

$$\frac{y}{y_0} = \frac{4x(L-x)}{8y_0 \frac{K}{n} t + L^2}$$

$$\frac{y_m}{y_0} = \frac{1}{8\frac{Ky_0 t}{nL^2} + 1}$$

$$L = \sqrt{\frac{8y_m \frac{Kt}{n}}{1 - \frac{y_m}{y_0}}} \qquad 8.21$$

The volume of water which flows out of the soil to one drain in time Δt is

$$\Delta \mathcal{V} = \int_0^L n \frac{\partial y}{\partial t} \Delta t \, dx$$

and the flow rate is

$$q = \frac{\Delta V}{\Delta t} = \int_0^L n \frac{\partial y}{\partial t} dx$$

$$= -\frac{\pi^2 K\bar{y}y_0}{L^2} \exp\left[-\left(\frac{\pi}{L}\right)^2 \frac{K\bar{y}}{n} t\right] \int_0^L \sin\frac{\pi x}{L} dx$$

$$= -2\pi \frac{K\bar{y}y_0}{L} \exp\left[-\left(\frac{\pi}{L}\right)^2 \frac{K\bar{y}}{n} t\right] \qquad 8.22$$

using equation 8.18. At $t = 0$

$$q = \frac{2\pi K\bar{y}y_0}{L} \qquad 8.23$$

Permeability may be estimated by testing soil samples in a permeameter in a laboratory. However, unless many such samples are tested, the permeabilities so obtained are local and may not be accurate for analysis of flows involving large volumes of the porous medium. This difficulty can be overcome to some extent by field testing.

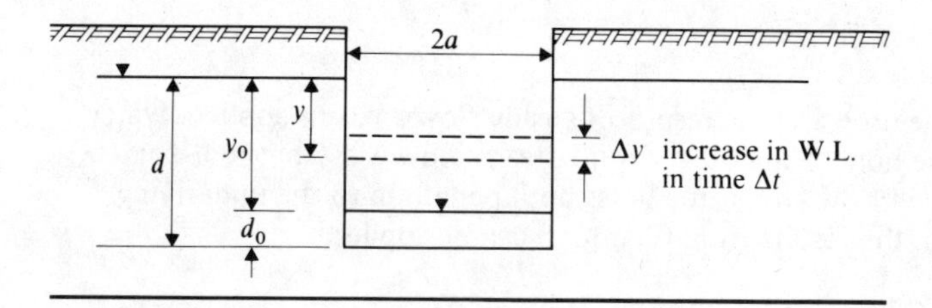

Fig. 8.5. Definition sketch for permeability test.

In the simplest field test, a single hole is dug or drilled with an auger, as sketched on Fig. 8.5. The water level in the hole is allowed to reach equilibrium with the water table. Then the hole is pumped out quickly and, after pumping is stopped, the rate at which the water level rises again is observed. From a simplified analysis of this unsteady flow the Hooghoudt formula is derived. It is assumed that the water table near the hole does not fall, that the flow through the wall of the hole is horizontal and that the flow through the bottom of the hole is vertical.

Then, the rate of flow through the wall is $(2\pi ad)Ky/S$ and that through the bottom is $(\pi a^2)Ky/S$. Here $2\pi ad$ and πa^2 are the relevant areas and y/S is the average slope of the hydraulic gradient. The length S depends on the lengths a,

d, z and y and is given by an empirical formula:

$$S = ad/0.19 \tag{8.24}$$

where S, a and d are measured in metres.

If ΔV is the volume of water which flows into the hole in time Δt

$$\Delta V = \Delta t(2\pi ad + \pi a^2)Ky/S$$

and

$$\Delta V = \pi a^2(-\Delta y)$$

Hence,

$$\frac{\mathrm{d}y}{\mathrm{d}t} = -K\frac{2d + a}{aS}y$$

and

$$\ln\frac{y_0}{y} = K(2d + a)\frac{t}{aS}$$

where $y = y_0$ at $t = 0$. Hence,

$$K = \frac{aS}{(2d + a)t}\ln\frac{y_0}{y} = \frac{2.3aS}{(2d + a)t}\log\frac{y_0}{y} \tag{8.25}$$

When a hole drilled to the impervious layer is used, water flows into the hole through the wall only. In that case

$$K = \frac{2.3aS}{2td}\log\frac{y_0}{y} \tag{8.26}$$

Two holes can be used and the results of steady flow analysis applied. Water is pumped from one hole to the other until a steady flow is established. This is the well-recharge well flow and, if the holes both penetrate to the underlying impervious stratum, the results from Chapter 3 can be applied:

$$K = \frac{Q\ln 2x_1/r_w}{\pi(z_0^2 - z_w^2)} = \frac{Q\ln 2x_1/r_w}{\pi s(2z_0 - s)} \simeq \frac{Q\ln 2x_1/r_w}{2\pi z_0 s} \tag{8.27}$$

where $2x_1$ is the distance between the holes, z_0 is the initial elevation of the water table and s is the drawdown at the well and the build up at the recharge well, that is the difference between the water levels in the holes, is $2s$. The last approximation above is valid when s is much smaller than z_0.

Unsteady two-dimensional flow can be used as an approximate description of the penetration of water from a lake or stream into the banks under the influence of changes in water level. For example, Fig. 8.6(*a*) shows schematically a lake with porous banks overlying an impermeable base. If the lake level is raised, a wave of groundwater penetrates the banks of the lake and may affect their stability.

This flow can be analysed approximately by using Dupuit's assumption, provided the increase in lake level is relatively small. Consideration of the flow

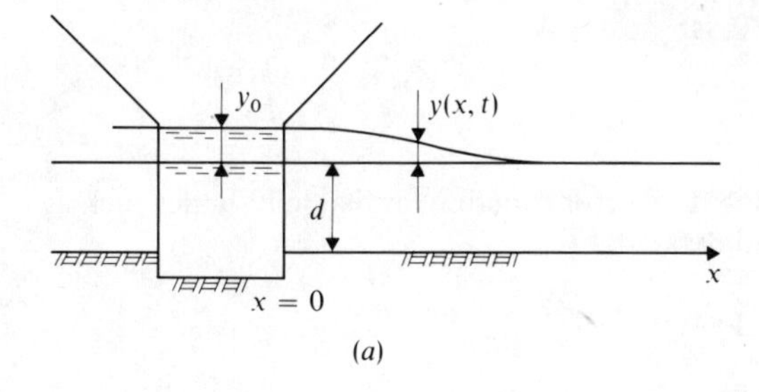

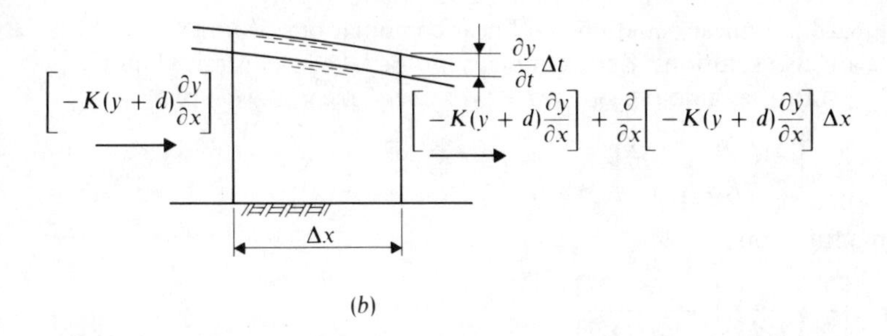

Fig. 8.6. Unsteady flow from a lake into its banks. (*a*) Definition sketch. (*b*) Elementary control volume.

to and from the control volume in Fig. 8.6(*b*) shows that equation 8.16 describes this flow also – that is

$$\frac{\partial}{\partial x}\left[K(y + d)\frac{\partial y}{\partial x}\right] = n\frac{\partial y}{\partial t} \qquad 8.16$$

where n is the porosity and y and d are defined on Fig. 8.6(*a*). We have assumed saturated pore space and constant permeability K. Attention is again drawn to the relevance of the diffusion equation, in this case, with diffusivity $K(y + d)/n$ dependent on the solution variable y.

The equation is readily solved if $y \ll d$, in which case

$$K(y + d) \simeq Kd$$

and equation 8.16 becomes

$$\frac{\partial y}{\partial t} = \frac{Kd}{n}\frac{\partial^2 y}{\partial x^2} \qquad 8.28$$

subject to the following boundary conditions:

$$x = 0, \quad t \leq 0, \quad y = 0$$

$$t > 0, \quad y = y_0 = \text{const}$$

$$x = \infty, \quad -\infty < t + \infty, \quad y = 0$$

The solution for $y(x, t), t > 0$, is

$$y = y_0 \operatorname{erfc}\left[\frac{1}{2}\left(\frac{n}{Kd}\right)^{1/2} \frac{x}{t^{1/2}}\right]$$

where erfc (s) is the complementary error function, available in tables such as Jahnke and Emde (1945) and defined by

$$\operatorname{erfc}(s) = 1 - \left(\frac{4}{\pi}\right)^{1/2} \int_0^s \exp(-\xi^2) d\xi$$

Thus, provided the properties K, d and n of the permeable banks are known, y can be found as a function of x and t for a given value of y_0.

Equation 8.28 is a linear equation and linear combinations of functions which are solutions are also solutions. For example, suppose lake level is raised by y_{01} at $t = 0$ and by a further amount y_{02} at $t = t_0$. For $t \leq t_0, y$ is given by

$$y = y_1 = y_{01} \operatorname{erfc}\left[\frac{1}{2}\left(\frac{n}{Kd}\right)^{1/2} \frac{x}{t^{1/2}}\right]$$

For $t > t_0$ the function

$$y_2 = y_{02} \operatorname{erfc}\left[\frac{1}{2}\left(\frac{n}{Kd}\right)^{1/2} \frac{x}{(t - t_0)^{1/2}}\right]$$

is superimposed to yield the linear combination

$$y = y_1 + y_2$$

The linearity of equation 8.28 also makes it possible to use image flows to satisfy some boundary conditions. Thus, if the impermeable stratum rises steeply at $x = x_0$, the region of flow can be represented approximately as shown in Fig. 8.7. When the flow from an image lake with its near bank at $x = 2x_0$ is superimposed on the flow from the real lake, the flow rate through the wall at $x = x_0$ is zero, as required. If the flow is to be investigated for a long period, the flow from the image may penetrate to $x = 0$, where a prescribed value of y must be maintained. This can be dealt with by means of a negative image at $x = -2x_0$; however, this is a less common problem.

The limitations of this analysis must not be forgotten. Equation 8.16 has been

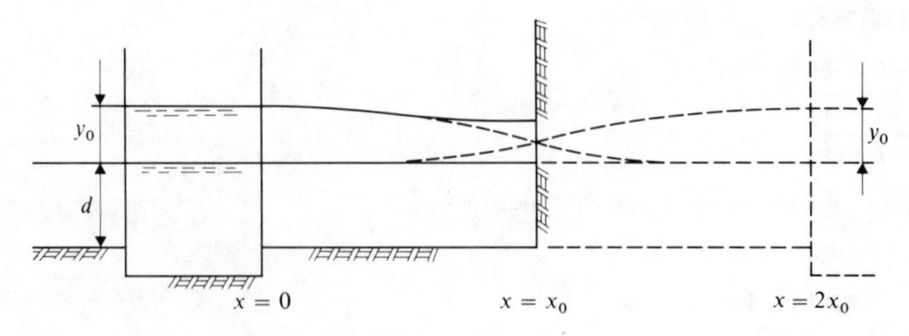

Fig. 8.7. Use of an image lake to simulate condition of no flow through $x = x_0$.

made amenable to closed solution by assuming the diffusivity to be constant. Thus, small values of y are implied, and, equally important, dependence of K on moisture content is excluded. No information is obtained about the advance of the capillary fringe above and ahead of the water table. All the limitations of the Dupuit assumptions are also relevant. Nevertheless, the analysis provides a useful simple estimate for flows in permeable strata which can be represented approximately as in Figs. 8.6 and 8.7 and for which the increase in lake level is small relative to the depth of water over the confining impermeable surface.

When the simplification required for the closed solution is not valid, individual problems must be solved by calculation or experiment. Philip (1969), Swartzendruber (1969) and Kirkham and Powers (1972) have reviewed the whole topic, and Kirkham and Powers give a detailed explanation of the methods of solution. Cedergren (1967) showed how graphical construction of flow nets can can be applied to the problem. The particular flow discussed above has been studied in detail by Cooper and Rorabaugh (1963) and their work is reviewed by Eagleson (1970).

Example 8.2

For the lake and its banks shown in Fig. 8.6(a), d = 10 m, $K = 10^{-6}$ m/s and $n = 0.4$. The water level in the lake is raised by 1.2 m and held at that level for 30 days. It is then lowered by 0.9 m.

Sketch the water table 30 days, 32 days and 40 days after the initial increase in the lake level.

$$\frac{1}{2}\sqrt{\frac{n}{K\mathrm{d}}} = \frac{1}{2}\sqrt{\frac{0.4}{10^{-6} \times 10}} = 100 \text{ m}^{-1}\text{s}^{1/2}$$

At $t = 30$ days $= 2.592 \times 10^6 s$

$$\frac{1}{2}\sqrt{\frac{n}{K\mathrm{d}}}\frac{x}{t^{1/2}} = 100\frac{x}{1610} = 0.0621\,x$$

$$y = 1.2\,\mathrm{erfc}(0.0621\,x)$$

$x =$	0	1	2	4	6	8	10
$0.0621\,x =$	0	0.0621	0.124	0.248	0.373	0.497	0.621
$y =$	1.2	1.116	1.033	0.871	0.718	0.579	0.456

$x =$	12	14
$0.0621\,x =$	0.745	0.870
$y =$	0.351	0.263

At $t = 32$ days, $t - t_0 = 2$ days $= 1.728 \times 10^5 s$

$$\frac{1}{2}\sqrt{\frac{n}{K\mathrm{d}}}\frac{x}{(t - t_0)^{1/2}} = 100\frac{x}{415.7} = 0.241\,x$$

$$y_1 = 1.2\,\mathrm{erfc}\,(0.0621\,x)$$

$$y_2 = -\,0.9\,\mathrm{erfc}\,(0.241\,x)$$

$x =$	0	1	2	4	6	8	10
$0.241x =$	0	0.241	0.481	0.962	1.443	1.925	2.406
$y_2 =$	−0.9	−0.660	−0.447	−0.137	−0.038	−0.006	−0.001
$y = y_1 + y_2 =$	0.3	0.456	0.586	0.734	0.680	0.573	0.455

$x =$	12	14
$0.241x =$	2.887	3.368
$y_2 =$	—	—
$y = y_1 + y_2 =$	0.351	0.263

At $t = 40$ days, $t - t_0 = 10$ days $= 8.64 \times 10^5 s$

$$\frac{1}{2}\sqrt{\frac{n}{Kd}}\frac{x}{(t - t_0)^{1/2}} = 100\frac{x}{929.5} = 0.1076\,x$$

$$y_1 = 1.2\,\text{erfc}(0.0621\,x)$$

$$y_2 = -0.9\,\text{erfc}(0.1076\,x)$$

$x =$	0	1	2	4	6	8	10
$0.1076x =$	0	0.1076	0.215	0.430	0.645	0.861	1.076
$y_2 =$	−0.9	−0.791	−0.685	−0.489	−0.326	−0.201	−0.116
$y = y_1 + y_2 =$	0.3	0.325	0.348	0.382	0.392	0.378	0.340

$x =$	12	14
$0.1076x =$	1.291	1.506
$y_2 =$	−0.061	−0.030
$y = y_1 + y_2 =$	0.290	0.233

The results of these calculations are shown on Fig. 8.8

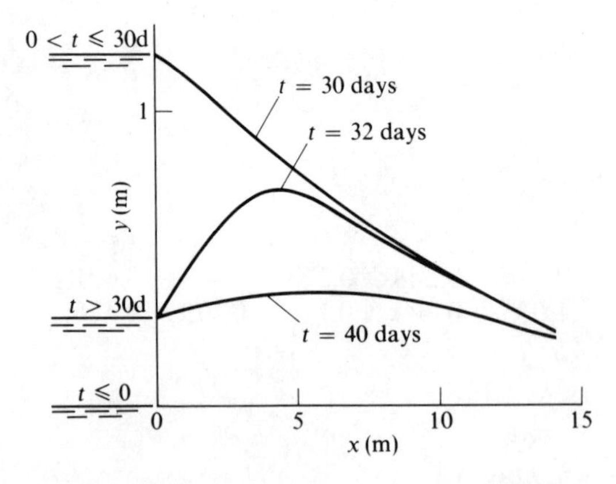

Fig. 8.8. Sketches of water table at various times for lake level raised at $t = 0$ and lowered at $t = 30$ days.

9
Salt water interfaces

Salt water interfaces

When a fresh water aquifer discharges into the sea fresh water and salt water meet. Mixing of the salt water and fresh water occurs, but only as a result of molecular diffusion; the most effective mechanism of mixing, large scale turbulence, does not occur in aquifers. Consequently, the boundary between salt water and fresh water in an aquifer is quite sharp and the thickness of the transition zone can be ignored. The aquifer can be divided into two regions, one occupied by fresh water, the other by salt water, the boundary between the two being known as the salt water interface.

Since the density of sea water is greater than that of fresh water, the former penetrates a coastal aquifer to some extent, underlying the lighter fresh water which flows above the sloping interface. The location of the interface is important in the development of groundwater resources since intrusion of sea water contaminates both the fresh water and the aquifer.

The essential features of the penetration of sea water into an aquifer are sketched in Fig. 9.1.

In the following notes properties of the interface are analysed for some simple flows, in most of which the salt water is at rest.

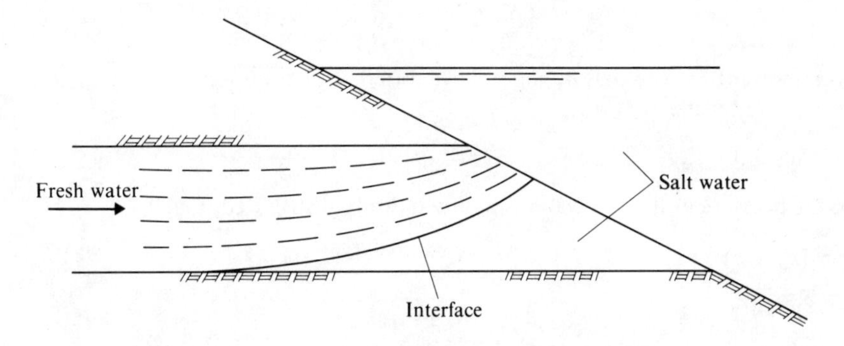

Fig. 9.1. Sketch showing salt water intrusion into an aquifer with fresh water flowing over the interface to the sea.

Flow of both liquids may be described by Darcy's law – in the region occupied by fresh water $V_f = -K_f \partial h_f / \partial s$ and in the region occupied by sea water $V_s = -K_s \partial h_f / \partial s$ where V_f and V_s are the velocities in the direction of s increasing, K_f and K_s are the transmission coefficients for fresh water and salt water in the aquifer and h_f and h_s are the piezometric heads in the two liquids. We have

$$K_f = \frac{k\gamma_f}{\mu_f} = \frac{kg}{\nu_f}$$

$$K_s = \frac{k\gamma_s}{\mu_s} = \frac{kg}{\nu_s}$$

$$h_f = \frac{p}{\gamma_f} + z$$

$$h_s = \frac{p}{\gamma_s} + z$$

It is necessary to distinguish between K_f and K_s because the transmission coefficient for the aquifer depends on the properties of the fluid as well as the properties of the aquifer (section 1.5). Equation 1.5 has been used to obtain the expressions above for K_f and K_s. It will be seen that, for a given aquifer they are inversely proportional to the kinematic viscosities of the liquids, so that $K_s : K_f \sim 0.96$. However, in aquifers with small pore size adsorption of ions on the surfaces of the soil particles can lead to a reduction of the effective pore size and consequently greater reduction in K_s.

Some attention must also be given to the piezometric heads. As defined above, and required for use in Darcy's law, the pressure head contribution in each case is the height of a column of water-fresh water in one region and salt water in the other. However, it may be more convenient to measure the pressure head in the region occupied by the salt water in terms of a column of fresh water. An alternative piezometric head h_s' for the region occupied by the salt water is then defined by

$$h_s' = \frac{p}{\gamma_f} + z$$

It is important to note that h_s and h_s' have different distributions. For example, when the salt water is at rest

$$h_s = \text{constant} = z_0$$

and the pressure in the salt water is hydrostatically distributed, Fig. 9.2,

$$p = \gamma_s(z_0 - z)$$

Consequently

$$h_s' = \frac{\gamma_s(z_0 - z)}{\gamma_f} + z = \frac{\gamma_s}{\gamma_f} z_0 - \frac{\Delta\gamma}{\gamma_f} z$$

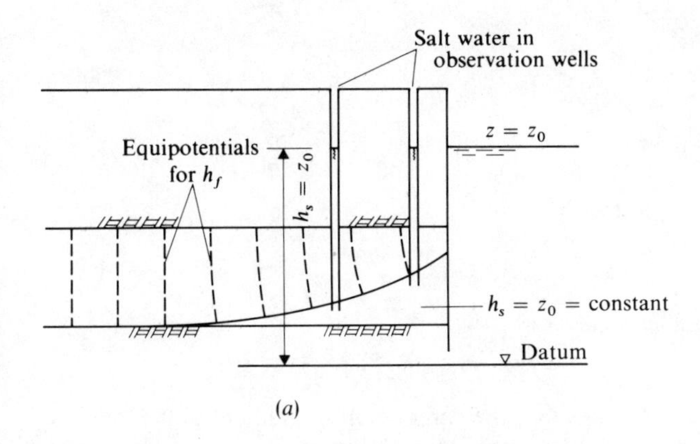

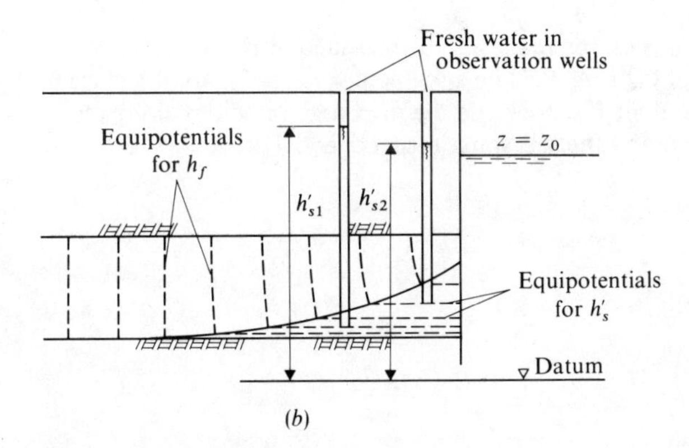

Fig. 9.2. The distribution of h_s and h'_s for stationary salt water. (*a*) $h_s(= p/\gamma_s + z)$ is constant. (*b*) $h'_s(= p/\gamma_f + z)$ is a function of z.

where

$$\Delta\gamma = \gamma_s - \gamma_f$$

That is, h'_s varies with z and increases linearly as z decreases. Equipotential surfaces in the region occupied by salt water are horizontal planes if the liquid in the observation wells is fresh water. Figures 9.2(*a*) and (*b*) illustrate the difference between the piezometric heads in this case. Note that the equipotentials showing h_f and h'_s are continuous in value (but not in slope) at the interface.

The interface is a boundary for both regions and both flows must satisfy the following boundary conditions at the interface:

(1) The pressure at the interface is the same in both liquids
(2) The component of discharge velocity normal to the interface is the same for both liquids and equals the velocity of the interface normal to itself.

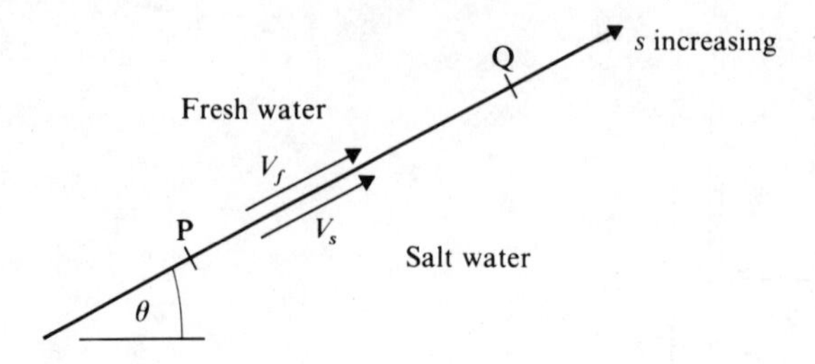

Fig. 9.3. Segment of a stationary interface.

From the first of these, it follows that the component of gradient of piezometric head tangential to the interface is the same in both liquids. The second condition has a corollary that the velocity in each liquid is tangential to the interface if the interface is stationary.

The following deductions can be made about the shape of the interface in a two-dimentional flow. Let PQ (Fig. 9.3) be a segment of a stationary interface in a two-dimensional flow and let V_f and V_s be the discharge velocities along the interface in the fresh water and the salt water respectively. Then

$$V_f = -K_f \frac{\partial h_f}{\partial s}$$

$$= -\frac{k\gamma_f}{\mu_f}\frac{\partial}{\partial s}\left(\frac{p_i}{\gamma_f} + z\right)$$

$$= -\frac{k}{\mu_f}\left(\frac{\partial p_i}{\partial s} + \gamma_f \frac{\partial z}{\partial s}\right)$$

$$V_s = -\frac{k}{\mu_s}\left(\frac{\partial p_i}{\partial s} + \gamma_s \frac{\partial z}{\partial s}\right)$$

Here z is the elevation of a point in the interface and p_i is the pressure at the same point. Eliminating the pressure gradient yields for the slope of the interface

$$\sin\theta = \frac{\partial z}{\partial s} = \frac{\mu_f V_f - \mu_s V_s}{k\Delta\gamma}$$

When $q_s = 0$

$$\sin\theta = \frac{\mu_f V_f}{k\Delta\gamma}$$

Therefore, the interface rises in the direction of flow and the slope increases as V_f increases. The interfaces sketched on Figs. 9.1 and 9.2 are consistent with these conclusions; V_f increases as the sea is approached because the salt water intrusion restricts the depth of the aquifer available for fresh water flow.

A simple analysis of the problem makes use of the Dupuit assumption (section 5.2) and the Ghyben-Herzberg approximation, (Badon-Ghyben (1888), Bear (1972), Herzberg (1901).) In Fig. 9.4 a two-dimensional confined aquifer is sketched and an observation well filled with fresh water is shown in contact with

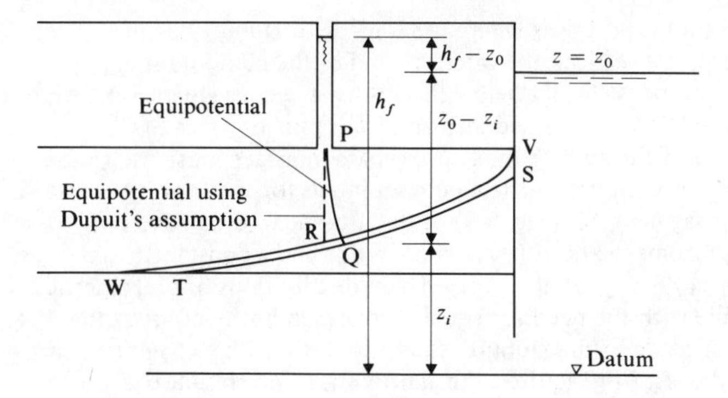

Fig. 9.4. Definition sketch for Ghyben-Herzberg approximation.

the aquifer where the piezometric head is h_f, measured with respect to an arbitrary horizontal datum. Two equipotentials are shown – one curved, PQ, being the true equipotential and one vertical, PR, being that assumed in the Dupuit approximation. Corresponding to these are the true interface and the one derived with the Dupuit assumption. If z_i is the elevation of Q in the true interface and R in the approximate one

$$h_f = \frac{p_i}{\gamma_f} + z_i$$

and p_i is given by

$$h_s = \frac{p_i}{\gamma_s} + z_i = z_0$$

since the salt water in the aquifer is at rest. Hence,

$$p_i = \gamma_s(z_0 - z_i)$$

and

$$h_f = \frac{\gamma_s}{\gamma_f}(z_0 - z_i) + z_i$$

From this we obtain

$$h_f - z_0 = \left(\frac{\gamma_s}{\gamma_f} - 1\right)(z_0 - z_i)$$

or

$$z_0 - z_i = \frac{\gamma_f}{\Delta\gamma}(h_f - z_0)$$

That is to say, the depth of the interface below sea level is $\gamma_f/\Delta\gamma$ times the height of the fresh water piezometric surface above sea level. If $\gamma_s/\gamma_f = 1.025$, this factor is 40.

Practical application is limited by the need to know the horizontal coordinate of the point Q and this is where use is made of Dupuit's assumption that the equipotentials are vertical; the interface end of the equipotential through P is assumed to be vertically below P. This has the consequence that the approximate interface VRW is obtained instead of the true interface SQT.

At the seaward end of the aquifer the approximate interface must rise to the point V to be consistent with the vertical equipotentials for the fresh water flow. If this were not so, a segment AB (Fig. 9.5) of the interface would have a constant sea water piezometric head ($= z_0$) on one side and a constant fresh water piezometric head ($= h_{f0}$) on the other. These distributions of piezometric head are incompatible with the need for equal pressures in both liquids at the interface. When use is made of the Dupuit assumption, B (in Fig. 9.5) must rise to A. At the seaward end of the aquifer, the approximate interface leaves no room for the fresh water to flow from the aquifer. It is clearly in error there and is correspondingly inaccurate over some length of the aquifer near the sea.

A similar analysis can be made for an unconfined aquifer. In that case, the phreatic surface and the interface meet at sea level on the coastline. In reality, they cannot; the interface is lower and the phreatic surface higher, with a seepage surface on the coast.

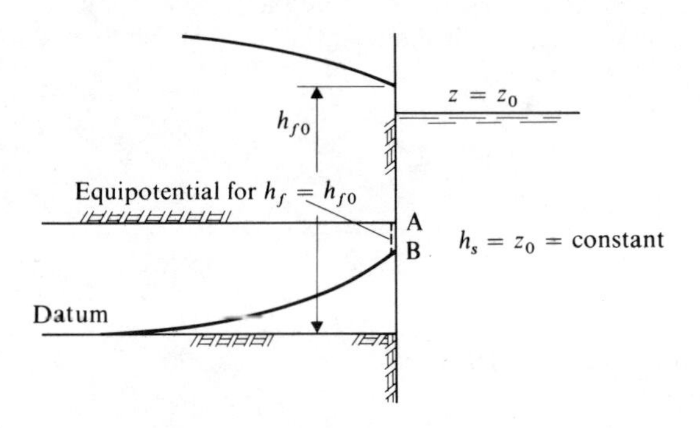

Fig. 9.5. Distributions of piezometric head at end of aquifer.

Notwithstanding the inaccuracy near the coast, the Dupuit-Ghyben-Herzberg approximation has been shown by Bear and Dagan (1962, see Bear 1972), to give the length of the salt water intrusion into a horizontal confined aquifer of thickness b to within 5% provided $\pi(\Delta\gamma/\gamma_f)K_f b/Q > 8$, where Q is the rate of flow of fresh water per unit breadth of the aquifer.

Application to some specific problems can now be undertaken. Figure 9.6 is a sketch of two-dimensional flow in a horizontal confined aquifer. The lower boundary of the aquifer is taken as datum for measurement of elevations and heads and the origin for x is at the end of the wedge of salt water. If Q is the rate of flow of fresh water per unit breadth of the aquifer and z_i is the elevation of the interface, then for $x > 0$

$$Q = -K \frac{dh_f}{dx} (b - z_i) \qquad 9.1$$

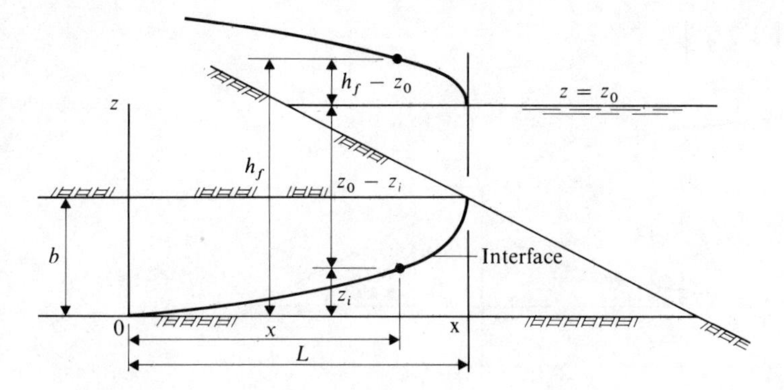

Fig. 9.6. Definition sketch for flow in a confined aquifer.

We also have

$$h_f - z_0 = \frac{\Delta\gamma}{\gamma_f}(z_0 - z_i)$$

whence

$$\frac{dh_f}{dx} = -\frac{\Delta\gamma}{\gamma_f}\frac{dz_i}{dx} \qquad 9.2$$

Equation 9.2 is used to eliminate h_f from equation 9.1 and

$$Q = K_f \frac{\Delta\gamma}{\gamma_f}\frac{dz_i}{dx}(b - z_i)$$

Hence,

$$Qx = K_f \frac{\Delta\gamma}{\gamma_f}(bz_i - \tfrac{1}{2}z_i^2) + \text{const}$$

Thus, the interface is a parabola with its vertex at $z_i = b$ and since it must pass through the points $x = 0$, $z_i = 0$ and $x = L$, $z_i = b$, we have

$$Qx = K_f \frac{\Delta\gamma}{\gamma_f}(bz_i - \tfrac{1}{2}z_i^2) \qquad 9.3$$

and

$$QL = \tfrac{1}{2}K_f \frac{\Delta\gamma}{\gamma_f} b^2$$

or

$$L = \frac{1}{2}\frac{\Delta\gamma}{\gamma_f}\frac{K_f b^2}{Q} \qquad 9.4$$

Another two-dimensional flow is shown on Fig. 9.7. Here the aquifer is unconfined and fresh water flows to the sea with the water table as its upper

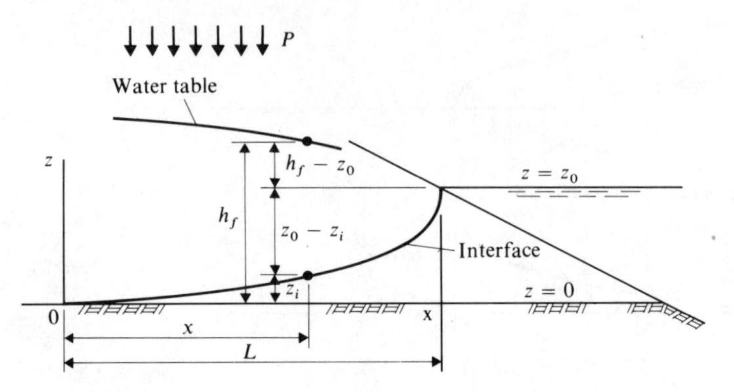

Fig. 9.7. Definition sketch for flow in an unconfined aquifer.

boundary and, where $x > 0$, the interface for its lower boundary. Including an allowance for precipitation at the rate P, the continuity equation yields

$$Q_0 + Px = -K_f \frac{dh_f}{dx} (h_f - z_i) \tag{9.5}$$

where Q_0 is the flow rate per unit breadth of the aquifer at $x = 0$. Using the Ghyben-Herzberg approximation

$$h_f - z_0 = \frac{\Delta\gamma}{\gamma_f} (z_0 - z_i)$$

so that

$$h_f - z_i = \frac{\gamma_s}{\gamma_f} (z_0 - z_i)$$

and

$$\frac{dh_f}{dx} = -\frac{\Delta\gamma}{\gamma_f} \frac{dz_i}{dx}$$

Hence,

$$Q_0 + Px = K_f \frac{\Delta\gamma}{\gamma_f} \frac{dz_i}{dx} \frac{\gamma_s}{\gamma_f} (z_0 - z_i)$$

$$Q_0 x + \tfrac{1}{2}Px^2 = K_f \frac{\Delta\gamma}{\gamma_f} \frac{\gamma_s}{\gamma_f} (z_0 z_i - \tfrac{1}{2}z_i^2) + \text{const}$$

The interface is again a parabola with vertex at $z_i = z_0$ and it must pass through $x = 0$, $z_i = 0$; $x = L$, $z_i = z_0$. It follows that

$$Q_0 x + \tfrac{1}{2}Px^2 = K_f \frac{\Delta\gamma}{\gamma_f} \frac{\gamma_s}{\gamma_f} (z_0 z_i - \tfrac{1}{2}z_i^2) \tag{9.6}$$

and

$$Q_0 L + \tfrac{1}{2}PL^2 = \tfrac{1}{2}K_f \frac{\Delta\gamma}{\gamma_f} \frac{\gamma_s}{\gamma_f} z_0^2 \tag{9.7}$$

When $P = 0$

$$L = \frac{1}{2}\frac{\Delta\gamma}{\gamma_f}\frac{\gamma_s}{\gamma_f}\frac{K_f z_0^2}{Q_0} \qquad 9.8$$

These are steady flows and equations 9.4, 9.7 and 9.8 can be used to determine the flow rate Q or Q_0 necessary to ensure that the sea does not intrude beyond a specified limit. Alternatively, they may be used to estimate the penetration to be expected when flow rates or precipitation or both are known.

Equations 9.4, 9.7 and 9.8 show that the position of the interface in a given aquifer is determined by the flow rate. When a change in flow occurs, the interface must be adjusted to conform with the new flow rate and one must ask how long this transition period will last. This is an unsteady flow problem. For the interface in a horizontal confined aquifer of thickness b a non-linear second order partial differential equation can be derived as follows (see Figs. 9.8(a) and (b)):

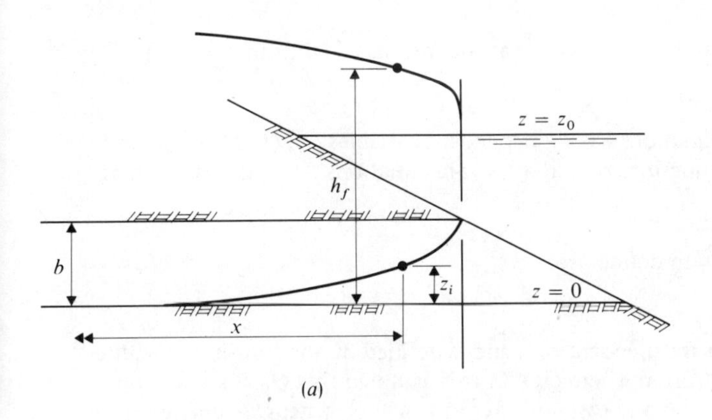

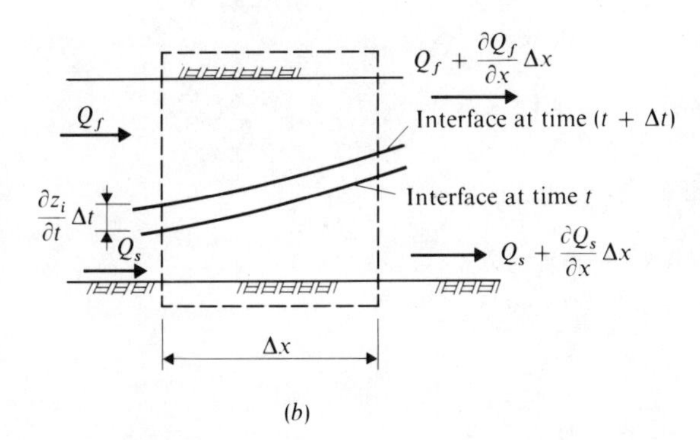

Fig. 9.8. Unsteady flow in a confined aquifer. (a) General scheme. (b) Details of flow through an elementary control volume.

The principle of conservation of matter applied separately to the fresh water and the salt water in a control volume yields

$$\frac{\partial Q_f}{\partial x} - n\frac{\partial z_i}{\partial t} = 0 \qquad 9.9$$

$$\frac{\partial Q_s}{\partial x} + n\frac{\partial z_i}{\partial t} = 0 \qquad 9.10$$

where Q_f and Q_s are the flow rates per unit breadth of the aquifer and n is the porosity of the aquifer. Note, that volume changes arising from changes in strain of the aquifer and the liquid have been ignored.

Using Dupuit's assumption and Darcy's law we have

$$Q_f = -K_f(b - z_i)\frac{\partial h_f}{\partial x} \qquad 9.11$$

$$Q_s = -K_s z_i \frac{\partial h_s}{\partial x} \qquad 9.12$$

and since the pressure in both liquids at the interface is the same

$$p_i = \gamma_f(h_f - z_i) = \gamma_s(h_s - z_i) \qquad 9.13$$

This set of six equations has six dependent variables Q_f, Q_s, h_f, h_s, p_i and z_i, all to be found as functions of x and t. From equations 9.9 and 9.10 we have

$$Q_f + Q_s = f(t)$$

and it will be useful to define

$$Q_0 = Q_f + Q_s \qquad 9.14$$

The flow rate Q_0 is independent of x and is defined by the flow in the aquifer upstream of the intrusion where $Q_s = 0$. It is assumed that Q_0 is a known function of time – in many cases of interest it will be a stepwise change at $t = 0$. From equation 9.13

$$z_i = \frac{\gamma_s}{\Delta\gamma}h_s - \frac{\gamma_f}{\Delta\gamma}h_f$$

whence, using equations 9.11 and 9.12

$$\frac{\partial z_i}{\partial x} = -\frac{Q_s}{K'_s z_i} + \frac{Q_f}{K'_f(b - z_i)}$$

where $K'_s = (\Delta\gamma/\gamma_s)K_s$ and $K'_f = (\Delta\gamma/\gamma_f)K_f$.

Substitution of

$$Q_f = Q_0 - Q_s$$

yields

$$\frac{\partial z_i}{\partial x} = -\frac{Q_s}{K'_s z_i} - \frac{Q_s}{K'_f(b - z_i)} + \frac{Q_0}{K'_f(b - z_i)}$$

and

$$Q_s = \frac{\dfrac{Q_0}{K_f'(b - z_i)} - \dfrac{\partial z_i}{\partial x}}{\dfrac{1}{K_s' z_i} + \dfrac{1}{K_f'(b - z_i)}}$$

Hence, using equation 9.10

$$n \frac{\partial z_i}{\partial t} + \frac{\partial}{\partial x}\left[\frac{\dfrac{Q_0}{K_f'(b - z_i)} - \dfrac{\partial z_i}{\partial x}}{\dfrac{1}{K_s' z_i} + \dfrac{1}{K_f'(b - z_i)}}\right] = 0 \qquad 9.15$$

Closed solutions for equation 9.15 are not available. The procedures and results for numerical solution are discussed by Bear (1972). The particular problem investigated by Bear is the advance of the interface when the flow rate Q_0 is suddenly reduced from one steady value to another smaller steady value. The interface for each of these flows is given by equations 9.3 and 9.4. The unsteady flow treated by Bear occurs as the interface is adjusted. In view of the difficulty of solving equation 9.15, we do not pursue this topic further.

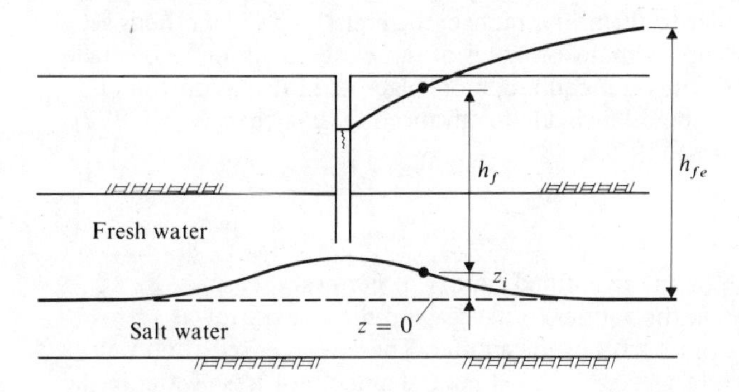

Fig. 9.9. Upconing accompanies pumping of fresh water from above a horizontal interface.

Problems of a different kind are encountered when fresh water is pumped from an aquifer in which the interface is horizontal, with the fresh water over overlying a body of stationary saline water. Then "upconing" occurs (Fig. 9.9). The interface rises in the vicinity of the well as the fresh water piezometric head decreases. If the pumping rate is high enough, the interface can rise sufficiently for salt water to be drawn into the well.

Complete solution of the problem is beyond our scope here, but an approximate estimate of the rise in level of the interface can be made for a small change (Muskat, 1937; Bear, 1972). The distribution of the fresh water piezometric head is estimated for confined aquifer with the undisturbed horizontal interface for its lowest boundary. Provided the upconing is small (strictly speaking, infinitesimal) it can be found on the basis of equal pressures in both liquids at the interface and constant salt water piezometric head in the stationary salt water, as follows:

$$p_i = \gamma_f(h_f - z_i)$$

$$h_s = \frac{\gamma_f(h_f - z_i)}{\gamma_s} + z_i$$

$$h_s = \frac{\gamma_f}{\gamma_s} h_f + \frac{\Delta\gamma}{\gamma_s} z_i = \text{const}$$

If $h_f = h_{fe}$ at a large distance where $z_i = 0$

$$\frac{\gamma_f}{\gamma_s} h_f + \frac{\Delta\gamma}{\gamma_s} z_i = \frac{\gamma_f}{\gamma_s} h_{fe}$$

and

$$z_i = \frac{\gamma_f}{\Delta\gamma}(h_{fe} - h_f) \qquad 9.16$$

It must be emphasized that this result is approximate and is valid only for a small rise in level of the interface.

Problems involving interfaces are complex and this brief survey does no more than introduce the reader to them. In practice, the equations and methods set out here should allow approximate estimates of salt water intrusion to be made. When more precise estimates are required, more elaborate calculation will be necessary and reference should be made to advanced texts such as Bear (1972) and Polubarinova-Kochina (1962).

Example 9.1

Figure 9.10 is a sketch of an unconfined aquifer of permeability $K_f = 9 \times 10^{-4}$ m/s. Fresh water flows in the aquifer to the sea and the rate of flow is 3×10^{-4} l/s per metre of breadth of the aquifer. There is an observation well 150 m from the coastline, as sketched. Estimate the depth of fresh water in the well.

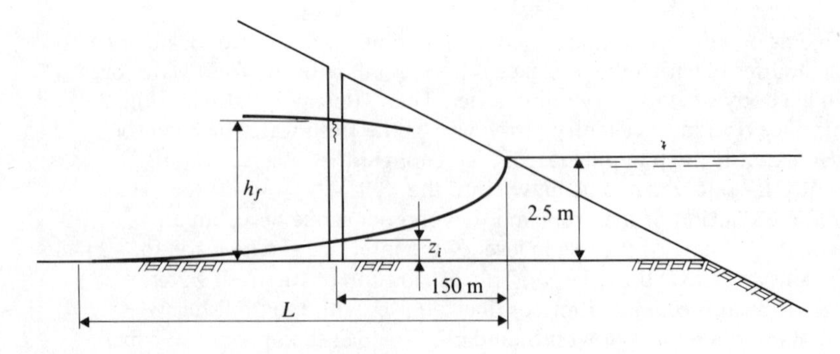

Fig. 9.10. Flow in an unconfined aquifer.

If water is pumped from the well at a low rate, so as to lower the water table at the well by 0.5 cm, what will happen to the interface?

The specific gravity of sea water is 1.025.

The length of the salt water intrusion is given by equation 9.8. Hence

$$L = \frac{1}{2} \frac{\Delta\gamma}{\gamma_f} \frac{\gamma_s}{\gamma_f} \frac{K_f z_0^2}{Q_0}$$

$$= \frac{1}{2} \times \frac{1}{40} \times 1.025 \times \frac{9 \times 10^{-4} \times 2.5 \times 2.5}{3 \times 10^{-7}}$$

$$= 240 \text{ m}$$

Thus, the origin of coordinates (Fig. 9.7) is 240 m from the coastline and we seek the levels of the water table and the interface where

$$x = 240 - 150 = 90 \text{ m}$$

From equation 9.6, with $P = 0$

$$z_i^2 - 2z_0 z_i = -\frac{2Q_0 x}{K_f} \frac{\gamma_f}{\Delta\gamma} \frac{\gamma_f}{\gamma_s}$$

$$= -\frac{2 \times 3 \times 10^{-7} \times 90}{9 \times 10^{-4}} \times 40 \times \frac{1}{1.025}$$

$$= -2.34 \text{ m}^2$$

$$(z_0 - z_i)^2 = -2.34 + 2.5^2 = 3.91 \text{ m}^2$$

$$z_0 - z_i = \pm 1.98 \text{ m}$$

$$z_i = 2.5 - 1.98 = 0.52 \text{ m}$$

Using the Ghyben-Herzberg approximation

$$h_f - z_0 = \frac{\Delta\gamma}{\gamma_f} (z_0 - z_i)$$

$$= \frac{1}{40} \times 1.98$$

$$= 0.0495 \text{ m}$$

$$h_f = 2.55 \text{ m}$$

Depth of fresh water in well $= h_f - z_i$

$$= 2.55 - 0.52$$

$$= 2.03 \text{ m}$$

Check that $\pi \frac{\Delta\gamma}{\gamma_f} \frac{K_f b}{Q} > 8$.

Taking $b = 2.5$ m, which is conservatively low

$$\pi \frac{\Delta\gamma}{\gamma_f} \frac{K_f b}{Q} = \pi \times \frac{1}{40} \times \frac{9 \times 10^{-4} \times 2.5}{3 \times 10^{-7}} = 590 \gg 8$$

If the water table is lowered by 0.5 cm equation 9.16 gives the amount by which the interface will rise:

$$h_{fe} - h_f = 5 \times 10^{-3} \text{ m}$$

$$\frac{\gamma_f}{\Delta\gamma}(h_{fe} - h_f) = 0.2 \text{ m}$$

i.e the interface will rise an additional 0.2 m in the well.

Example 9.2

A horizontal confined aquifer, Fig. 9.11(a), whose thickness is 13 m and permeability with respect to fresh water 0.07 cm/s discharges water into the sea at the rate of 8.25×10^{-3} l/s per metre of breadth. The aquifer is to be developed as a source of fresh water by pumping from a line of wells parallel to

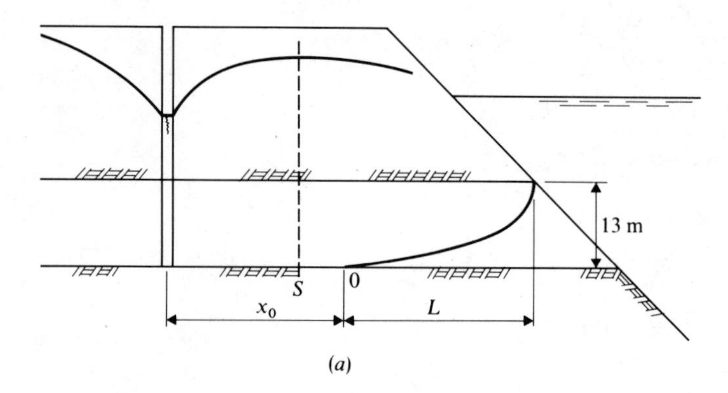

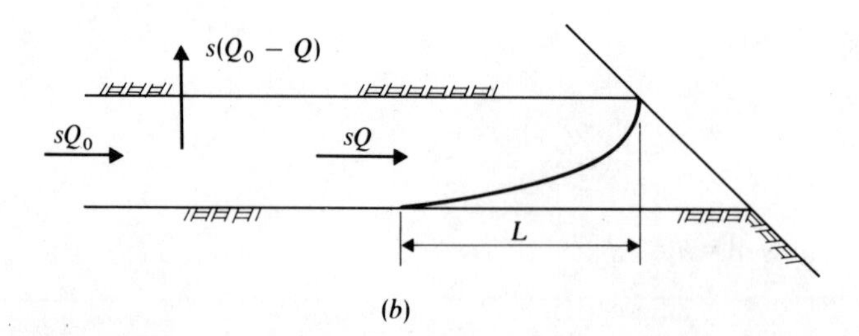

Fig. 9.11. A line of wells near a coastline. (a) Vertical section through one well. (b) Flow rates in an aquifer and one well. (c) Plan view of wells and limit of interface, with improved estimate of velocity distribution downstream of well. (d) Bounding streamline for well in uniform flow. (e) Bounding streamline for well, uniform flow and image well.

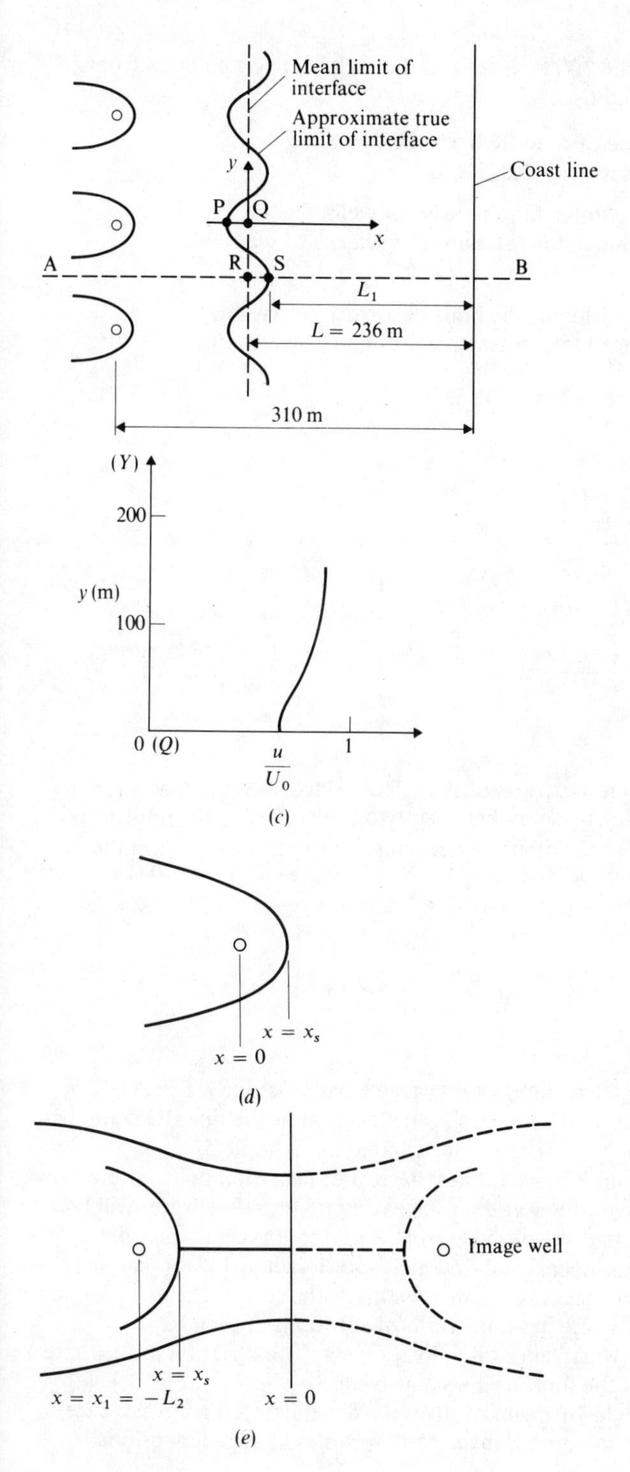
Mean limit of interface
Approximate true limit of interface
Coast line
y
x
P
Q
R
S
A
B
L_1
$L = 236$ m
310 m
(Y)
200
y (m)
100
0 (Q)
$\frac{u}{U_0}$
1
$x = x_s$
$x = 0$
Image well
$x = x_s$
$x = x_1 = -L_2$
$x = 0$

(c)

(d)

(e)

the coast. The wells will be 300 m apart and the water is to be pumped from each well at the rate of 0.6 l/s.

How far should the line of wells be from the coastline?
Specific gravity of sea water = 1.025.

Let Q_0 = rate of flow in aquifer far upstream of wells
Q = average rate of flow downstream of influence of wells
s = spacing of wells

(1) Estimate the mean position of the limit of intrusion. Assume two-dimensional flow at reduced rate downstream of line of wells, Fig. 9.11(*b*):

$$Q_0 = 8.25 \times 10^{-6} \text{ m}^2/\text{s}$$

$$s(Q_0 - Q) = 6 \times 10^{-4} \text{ m}^3/\text{s}$$

$$s = 300 \text{ m}$$

$$Q_0 - Q = 2 \times 10^{-6} \text{ m}^2/\text{s}$$

$$Q = 6.25 \times 10^{-6} \text{ m}^2/\text{s}$$

Substitution in equation 9.4 yields

$$L = \frac{1}{2} \times \frac{1}{40} \times \frac{7 \times 10^{-4} \times 13 \times 13}{6.25 \times 10^{-6}}$$

$$= 236 \text{ m}$$

(2) Estimate the maximum intrusion of the saline wedge. Assume that (*a*) along a line like AB, Fig. 9.11(*c*), half way between two wells, $Q = Q_0$, the undisturbed rate of flow (*b*) PQ = RS – i.e. that the departure of the true limit from the mean is symmetrical

$$L_1 = \frac{6.25}{8.25} \times 236$$

$$= 179 \text{ m}$$

$$\text{RS} = 236 - 179 = 57 \text{ m}$$

Distance of estimated extreme limit of interface from coastline = L + PQ = L + RS = 236 + 57 = 293 m. Alternatively, the location of the line RQ could be estimated from the velocity distribution calculated in Example 3.2.
(3) Estimate distance from P to well. Let P be at the stagnation point of the flow to the well. If P is closer to the well than stagnation point, saline water will be drawn into the well. To find the distance from a well to its stagnation point requires analysis of three-dimensional flow fields for individual wells, because each well in the row influences every other well (c.f. the row of drains in Section 3.8) and because the flow rate per unit width of aquifer is uneven for a considerable distance downstream of the line of wells. The variation in flow rate causes the wavy form of the limit of the saline wedge as shown in Fig. 9.11(*c*) and this makes the flow downstream of the wells dependent on all three space coordinates. In the next step, use is made of the studies of two-dimensional

fields in Chapter 3. Effects of overlapping of flow fields are neglected, and three alternatives, each with a single well, are used as approximations for the three-dimensional field. We assume that
(a) the wells are far enough apart for analysis of a single well to be a good approximation
(b) that we have a well in a uniform flow or
(c) that we have the well as in (b) with a recharge well symmetrically placed with respect to RQ, or
(d) that we have a well and a recharge well as in (c), except that the line of symmetry for the well and the recharge well is the coastline.

Model (b)

$$x_s = \frac{q}{2\pi U_0} \qquad 3.31$$

$$= \frac{s(Q_0 - Q)}{2\pi Q_0}$$

$$= \frac{1}{2\pi}\left(1 - \frac{Q}{Q_0}\right) s$$

$$= \frac{1}{2\pi}\left(1 - \frac{6.25}{8.25}\right) \times 300$$

$$= 11.6 \text{ m}$$

Total distance of well from coastline = 293 + 11.6 = 305 m

Model (c)

$$x_s = x_1 \sqrt{1 + \frac{q}{\pi x_1 U_0}} \qquad 3.51$$

$$= -L_2 \sqrt{1 - \frac{s(Q_0 - Q)}{\pi L_2 Q_0}}$$

$$= -L_2 \sqrt{1 - \frac{1}{\pi}\left(1 - \frac{Q}{Q_0}\right)\frac{s}{L_2}}$$

Setting x_s = RS

$$x_s = 57 \text{ m}$$

$$57^2 = L_2^2\left(1 - \frac{23.2}{L_2}\right) = L_2^2 - 23.2\,L_2$$

$$(L_2 - 11.6)^2 = 57^2 + 11.6^2 = 58.2^2$$

$$L_2 = 69.8 \text{ m}$$

Total distance of well from coastline = 236 + 69.8 = 306 m

Model (d)

As for model (c), but with $x_s = -293$ m

In this case, L_2 is the total distance of the well from the coastline and is found to be 305 m.

Estimated safe position of line of wells is 310 m from coastline.

It must be emphasized that prudence would require location of the wells substantially further from the sea, unless this result is confirmed by more sophisticated analysis or test drilling and pumping. If the intrusion were to reach a well, equation 9.16 shows that upconing would occur and would be of the order of 40 times the drawdown at the well. Even an apparently minor intrusion under a well would have serious consequences.

Note, also the need to allow a substantial proportion of the total flow in the aquifer to pass the line of wells. The wells rely on this flow to arrest the salt water wedge.

A check calculation can be made by superimposing the flow fields for three adjacent wells. For example, using model (c) and Example 3.2 the velocity downstream of one of the wells would be distributed along RQ as follows:

$$\frac{u}{U_0} = 1 + \frac{q}{\pi x_1 U_0} \frac{1}{1 + (y/x_1)^2}$$

$$x_1 = -(310 - 236) = -74 \text{ m}$$

$$\frac{q}{\pi x_1 U_0} = \frac{(qb)}{\pi x_1 (U_0 b)}$$

$$= -\frac{6 \times 10^{-4}}{\pi \times 74 \times 8.25 \times 10^{-6}}$$

$$= -0.313$$

$$\frac{u}{U_0} = 1 - 0.313 \frac{1}{1 + (y/74)^2}$$

$y =$	0	50	100	150	200	250
$1 - \frac{u}{U_0} =$	0.313	0.215	0.111	0.0612	0.0377	0.0252

$y =$	300	350	400	450
$1 - \frac{u}{U_0} =$	0.0179	0.0134	0.0104	0.0082

If the origin of coordinates is now taken at Q (Fig. 9.11(*c*) with the x and y axes as shown on Fig. 9.11(*c*), the distribution of velocity for $0 \leq y \leq 150$ m is found as follows:

	$y = 0$	50	100	150
Undisturbed flow	1	1	1	1
Disturbance due to central well	–0.313	–0.215	–0.111	–0.0612
Disturbance due to well at y = 300 m	–0.0179	–0.0252	–0.0377	–0.0612
Disturbance due to well at y = –300 m	–0.0179	–0.0134	–0.0104	–0.0082
$\frac{u}{U_0}$	= 0.651	0.746	0.841	0.869

This distribution is shown on Fig. 9.11(*c*)
Hence, at Q

$$\frac{Q}{Q_0} = \frac{u}{U_0}\bigg|_{y=0} = 0.651$$

$$Q = 0.651 \times 8.25 \times 10^{-6}$$

$$= 5.371 \times 10^{-6}\ \text{m}^2/\text{s}$$

From equation 9.4

$$L = \frac{1}{2} \times \frac{1}{40} \times \frac{7 \times 10^{-4} \times 13 \times 13}{5.371 \times 10^{-6}}$$

$$= 275.33\ \text{m}$$

Thus, a better estimate of PQ is 275.33 – 236 or 40 m. This may be compared with 57 m estimated from RS plus 4 m in rounding off. The estimated position of the wells is seen to allow a margin of about 20 m, when compared to the estimate from this improved approximation.

Concluding note

A limited range of groundwater flows has been discussed. The emphasis has been on derivation, on showing how the analysis is built up. We have restricted ourselves to mainly simple problems, although here and there a few remarks or formulae have been given which venture further. The fluid has been assumed to be homogeneous, the aquifer has generally been infinite or semi-infinite and wells have been assumed to penetrate the entire thickness of aquifers.

Apart from problems related to the petroleum industry, the assumption of homogeneity of fluid is usually correct. Where the exclusion of polluted water is being investigated, the fluid is certainly not homogeneous in respect of its quality, but this does not make the assumption invalid for an investigation of the flow. Temperature variations can affect fluid properties and cause convection currents. This is particularly important in regions of volcanic activity and in relation to the development of geothermal power (Wooding (1963, 1964), Elder (1965)).

The shape, in plan, of the aquifer can be taken into account, at least in principle, by judicious placing of sinks and sources and the use of the concept of images. If the vertical spacing of the confining boundaries varies then z has to be introduced into equation 2.19 as a function of x and y and this may lead to a very complex analytical relationship which can be solved only in particular cases by numerical analysis.

A large number of numerical, graphical and analogue techniques have been exploited for solution of groundwater problems – the flow net and finite difference calculations, physical models such as sand tanks, analogue models like the Hele-Shaw apparatus, electrical resistance-capacitance networks, electrical resistance networks, electrolytic tank and conductive sheet. For details of these, the literature should be consulted, in particular Karplus (1958), De Wiest (1965) and Bear (1972).

We have not dealt with any of the aspects of investigation of groundwater resources, not because it is unimportant but because it is a topic in its own right. The investigations of aquifer geometry, properties, regions of recharge, etc. are difficult and employ a great variety of techniques, such as electric resistance measurements, seismic surveys, magnetic surveys, drilling, use of isotope tracers, study of the variation of tritium and oxygen–18 content. The International Atomic Energy Agency, Vienna, has published extensively in the field of use of isotopes in groundwater surveys. Other aspects are covered in literature on petroleum geology.

Attention is drawn to Huisman (1972), Verruijt (1970) and Walton (1970). In particular, Huisman and Walton present a great deal of information about the practical aspects of construction for the development of groundwater resources, such as aquifer tests, well-design criteria, groundwater exploration and quality, groundwater management.

Appendix I: formulae for potential flows

Formulae for Potential Flows

1 Uniform flow parallel to Ox in direction of x decreasing:

$$\phi = -U_0 x = -U_0 r \cos\theta$$

$$\psi = -U_0 y = -U_0 r \sin\theta$$

2 Radial flow to a sink at the origin of coordinates:

$$\phi = m \ln r = \frac{1}{2} m \ln(x^2 + y^2)$$

$$\psi = m\theta = m \tan^{-1}\left(\frac{y}{x}\right)$$

Strength of the sink $m = \dfrac{q}{2\pi} = \dfrac{Q}{2\pi b}$

Q = discharge at sink

b = thickness of aquifer

$q = \dfrac{Q}{b}$ = discharge per unit thickness of aquifer

3 Radial flow from a source at the origin of coordinates:

$$\phi = -m \ln r = -\frac{1}{2} m \ln(x^2 + y^2)$$

$$\psi = -m\theta = -m \tan^{-1}\left(\frac{y}{x}\right)$$

Strength of source m defined as in 2.

4 Source-sink pair:

Coordinates of source: $(-x_1, 0)$

Coordinates of sink: $(x_1, 0)$

$$\phi = m \ln \frac{r_1}{r_2} = \frac{1}{2} m \ln \frac{(x - x_1)^2 + y^2}{(x - x_1)^2 + y^2}$$

$$\psi = m(\theta_1 - \theta_2) = m \left[\tan^{-1} \left(\frac{y}{x - x_1} \right) - \tan^{-1} \left(\frac{y}{x + x_1} \right) \right]$$

Source and sink are of equal strength m and $m = q/2\pi$ as above. r_1, θ_1 are the coordinates of (x, y) referred to the sink and Ox.

r_2, θ_2 are the coordinates of (x, y) referred to the source and Ox.

Appendix II

$\frac{r}{\sqrt{4Tt}}$	$\int_{r/\sqrt{4Tt}}^{\infty}\frac{1}{u}e^{-u^2}du$	$\frac{r}{\sqrt{4Tt}}$	$\int_{r/\sqrt{4Tt}}^{\infty}\frac{1}{u}e^{-u^2}du$	$\frac{r}{\sqrt{4Tt}}$	$\int_{r/\sqrt{4Tt}}^{\infty}\frac{1}{u}e^{-u^2}du$	$\frac{r}{\sqrt{4Tt}}$	$\int_{r/\sqrt{4Tt}}^{\infty}\frac{1}{u}e^{-u^2}du$
0.00010	8.92173	0.00084	6.79350	0.00680	4.70225	0.05200	2.66925
0.00011	8.82642	0.00085	6.78167	0.00690	4.68765	0.05300	2.65026
0.00012	8.73941	0.00086	6.76997	0.00700	4.67326	0.05400	2.63162
0.00013	8.65937	0.00087	6.75841	0.00710	4.65908	0.05500	2.61333
0.00014	8.58526	0.00088	6.74698	0.00720	4.64509	0.05600	2.59536
0.00015	8.51627	0.00089	6.73568	0.00730	4.63130	0.05700	2.57772
0.00016	8.45173	0.00090	6.72451	0.00740	4.61769	0.05800	2.56038
0.00017	8.39110	0.00091	6.71346	0.00750	4.60427	0.05900	2.54335
0.00018	8.33395	0.00092	6.70253	0.00760	4.59103	0.06000	2.52660
0.00019	8.27988	0.00093	6.69172	0.00770	4.57796	0.06100	2.51013
0.00020	8.22859	0.00094	6.68102	0.00780	4.56505	0.06200	2.49393
0.00021	8.17980	0.00095	6.67044	0.00790	4.55232	0.06300	2.47800
0.00022	8.13327	0.00096	6.65997	0.00800	4.53974	0.06400	2.46231
0.00023	8.08882	0.00097	6.64961	0.00810	4.52732	0.06500	2.44687
0.00024	8.04626	0.00098	6.63935	0.00820	4.51505	0.06600	2.43167
0.00025	8.00544	0.00099	6.62920	0.00830	4.50293	0.06700	2.41670
0.00026	7.96622	0.00100	6.61915	0.00840	4.49095	0.06800	2.40195
0.00027	7.92848	0.00110	6.52384	0.00850	4.47912	0.06900	2.38742
0.00028	7.89211	0.00120	6.43683	0.00860	4.46742	0.07000	2.37310
0.00029	7.85702	0.00130	6.35678	0.00870	4.45586	0.07100	2.35898
0.00030	7.82312	0.00140	6.28268	0.00880	4.44443	0.07200	2.34507
0.00031	7.79033	0.00150	6.21368	0.00890	4.43314	0.07300	2.33135
0.00032	7.75858	0.00160	6.14914	0.00900	4.42196	0.07400	2.31782
0.00033	7.72781	0.00170	6.08852	0.00910	4.41091	0.07500	2.30447
0.00034	7.69796	0.00180	6.03136	0.00920	4.39999	0.07600	2.29130
0.00035	7.66897	0.00190	5.97730	0.00930	4.38918	0.07700	2.27830
0.00036	7.64080	0.00200	5.92600	0.00940	4.37848	0.07800	2.26548
0.00037	7.61340	0.00210	5.87721	0.00950	4.36790	0.07900	2.25281
0.00038	7.58673	0.00220	5.83069	0.00960	4.35743	0.08000	2.24032
0.00039	7.56076	0.00230	5.78624	0.00970	4.34707	0.08100	2.22797
0.00040	7.53544	0.00240	5.74368	0.00980	4.33681	0.08200	2.21578
0.00041	7.51075	0.00250	5.70286	0.00990	4.32666	0.08300	2.20375
0.00042	7.48665	0.00260	5.66364	0.01000	4.31661	0.08400	2.19185
0.00043	7.46312	0.00270	5.62590	0.01100	4.22131	0.08500	2.18010
0.00044	7.44013	0.00280	5.58953	0.01200	4.13431	0.08600	2.16849
0.00045	7.41765	0.00290	5.55444	0.01300	4.05428	0.08700	2.15702
0.00046	7.39568	0.00300	5.52054	0.01400	3.98019	0.08800	2.14567
0.00047	7.37417	0.00310	5.48775	0.01500	3.91121	0.08900	2.13446
0.00048	7.35312	0.00320	5.45600	0.01600	3.84669	0.09000	2.12338
0.00049	7.33250	0.00330	5.42523	0.01700	3.78608	0.09100	2.11242
0.00050	7.31229	0.00340	5.39538	0.01800	3.72894	0.09200	2.10158
0.00051	7.29249	0.00350	5.36639	0.01900	3.67489	0.09300	2.09086
0.00052	7.27307	0.00360	5.33822	0.02000	3.62361	0.09400	2.08026
0.00053	7.25403	0.00370	5.31082	0.02100	3.57485	0.09500	2.06977
0.00054	7.23533	0.00380	5.28415	0.02200	3.52835	0.09600	2.05940
0.00055	7.21698	0.00390	5.25818	0.02300	3.48392	0.09700	2.04913
0.00056	7.19897	0.00400	5.23286	0.02400	3.44138	0.09800	2.03897
0.00057	7.18127	0.00410	5.20817	0.02500	3.40058	0.09900	2.02892
0.00058	7.16387	0.00420	5.18407	0.02600	3.36139	0.10000	2.01896
0.00059	7.14678	0.00430	5.16054	0.02700	3.32367	0.10100	2.00911
0.00060	7.12997	0.00440	5.13755	0.02800	3.28733	0.10200	1.99936
0.00061	7.11344	0.00450	5.11508	0.02900	3.25227	0.10300	1.98971
0.00062	7.09718	0.00460	5.09310	0.03000	3.21840	0.10400	1.98015
0.00063	7.08118	0.00470	5.07160	0.03100	3.18564	0.10500	1.97068
0.00064	7.06543	0.00480	5.05054	0.03200	3.15392	0.10600	1.96131
0.00065	7.04993	0.00490	5.02992	0.03300	3.12318	0.10700	1.95203
0.00066	7.03466	0.00500	5.00972	0.03400	3.09336	0.10800	1.94283
0.00067	7.01962	0.00510	4.98992	0.03500	3.06441	0.10900	1.93372
0.00068	7.00481	0.00520	4.97050	0.03600	3.03628	0.11000	1.92470
0.00069	6.99021	0.00530	4.95145	0.03700	3.00891	0.11100	1.91576
0.00070	6.97582	0.00540	4.93276	0.03800	2.98228	0.11200	1.90690
0.00071	6.96164	0.00550	4.91441	0.03900	2.95635	0.11300	1.89812
0.00072	6.94765	0.00560	4.89640	0.04000	2.93107	0.11400	1.88943
0.00073	6.93386	0.00570	4.87870	0.04100	2.90642	0.11500	1.88081
0.00074	6.92025	0.00580	4.86131	0.04200	2.88236	0.11600	1.87226
0.00075	6.90683	0.00590	4.84421	0.04300	2.85887	0.11700	1.86379
0.00076	6.89358	0.00600	4.82741	0.04400	2.83593	0.11800	1.85540
0.00077	6.88051	0.00610	4.81088	0.04500	2.81350	0.11900	1.84708
0.00078	6.86761	0.00620	4.79462	0.04600	2.79156	0.12000	1.83883
0.00079	6.85487	0.00630	4.77862	0.04700	2.77010	0.12100	1.83065
0.00080	6.84229	0.00640	4.76287	0.04800	2.74910	0.12200	1.82254
0.00081	6.82987	0.00650	4.74737	0.04900	2.72853	0.12300	1.81450
0.00082	6.81760	0.00660	4.73210	0.05000	2.70837	0.12400	1.80652
0.00083	6.80548	0.00670	4.71706	0.05100	2.68862	0.12500	1.79862

Values of the Integral $\int_{r/\sqrt{4Tt}}^{\infty} \frac{1}{u} e^{-u^2}\, du$ for Given Values of the Parameter $r/\sqrt{4Tt}$. (Glover, U. S. Bureau of Reclamation, Engineering Monograph No. 31).

$\frac{r}{\sqrt{4Tt}}$	$\int_{r/\sqrt{4Tt}}^{\infty} \frac{1}{u} e^{-u^2} du$	$\frac{r}{\sqrt{4Tt}}$	$\int_{r/\sqrt{4Tt}}^{\infty} \frac{1}{u} e^{-u^2} du$	$\frac{r}{\sqrt{4Tt}}$	$\int_{r/\sqrt{4Tt}}^{\infty} \frac{1}{u} e^{-u^2} du$	$\frac{r}{\sqrt{4Tt}}$	$\int_{r/\sqrt{4Tt}}^{\infty} \frac{1}{u} e^{-u^2} du$
0.12600	1.79077	0.20800	1.30301	0.29000	.99045	0.37200	.76712
0.12700	1.78299	0.20900	1.29842	0.29100	.98728	0.37300	.76479
0.12800	1.77528	0.21000	1.29385	0.29200	.98413	0.37400	.76246
0.12900	1.76762	0.21100	1.28930	0.29300	.98100	0.37500	.76014
0.13000	1.76003	0.21200	1.28478	0.29400	.97787	0.37600	.75782
0.13100	1.75249	0.21300	1.28028	0.29500	.97476	0.37700	.75552
0.13200	1.74502	0.21400	1.27581	0.29600	.97165	0.37800	.75322
0.13300	1.73760	0.21500	1.27136	0.29700	.96857	0.37900	.75093
0.13400	1.73025	0.21600	1.26693	0.29800	.96549	0.38000	.74865
0.13500	1.72294	0.21700	1.26252	0.29900	.96242	0.38100	.74638
0.13600	1.71570	0.21800	1.25813	0.30000	.95937	0.38200	.74411
0.13700	1.70851	0.21900	1.25377	0.30100	.95633	0.38300	.74185
0.13800	1.70137	0.22000	1.24943	0.30200	.95330	0.38400	.73960
0.13900	1.69429	0.22100	1.24511	0.30300	.95029	0.38500	.73736
0.14000	1.68726	0.22200	1.24081	0.30400	.94728	0.38600	.73512
0.14100	1.68028	0.22300	1.23653	0.30500	.94429	0.38700	.73289
0.14200	1.67335	0.22400	1.23228	0.30600	.94131	0.38800	.73067
0.14300	1.66648	0.22500	1.22804	0.30700	.93834	0.38900	.72846
0.14400	1.65965	0.22600	1.22383	0.30800	.93538	0.39000	.72625
0.14500	1.65287	0.22700	1.21963	0.30900	.93243	0.39100	.72406
0.14600	1.64614	0.22800	1.21546	0.31000	.92949	0.39200	.72186
0.14700	1.63946	0.22900	1.21131	0.31100	.92657	0.39300	.71968
0.14800	1.63283	0.23000	1.20717	0.31200	.92366	0.39400	.71750
0.14900	1.62624	0.23100	1.20306	0.31300	.92075	0.39500	.71533
0.15000	1.61970	0.23200	1.19896	0.31400	.91786	0.39600	.71317
0.15100	1.61320	0.23300	1.19489	0.31500	.91498	0.39700	.71102
0.15200	1.60675	0.23400	1.19083	0.31600	.91211	0.39800	.70887
0.15300	1.60035	0.23500	1.18680	0.31700	.90926	0.39900	.70673
0.15400	1.59398	0.23600	1.18278	0.31800	.90641	0.40000	.70459
0.15500	1.58766	0.23700	1.17878	0.31900	.90357	0.40100	.70247
0.15600	1.58139	0.23800	1.17480	0.32000	.90074	0.40200	.70035
0.15700	1.57515	0.23900	1.17084	0.32100	.89793	0.40300	.69823
0.15800	1.56896	0.24000	1.16690	0.32200	.89512	0.40400	.69613
0.15900	1.56280	0.24100	1.16297	0.32300	.89233	0.40500	.69403
0.16000	1.55669	0.24200	1.15907	0.32400	.88955	0.40600	.69194
0.16100	1.55062	0.24300	1.15518	0.32500	.88677	0.40700	.68985
0.16200	1.54459	0.24400	1.15131	0.32600	.88401	0.40800	.68777
0.16300	1.53859	0.24500	1.14746	0.32700	.88126	0.40900	.68570
0.16400	1.53264	0.24600	1.14362	0.32800	.87851	0.41000	.68364
0.16500	1.52672	0.24700	1.13980	0.32900	.87578	0.41100	.68158
0.16600	1.52084	0.24800	1.13600	0.33000	.87306	0.41200	.67953
0.16700	1.51500	0.24900	1.13222	0.33100	.87034	0.41300	.67748
0.16800	1.50920	0.25000	1.12845	0.33200	.86764	0.41400	.67544
0.16900	1.50343	0.25100	1.12471	0.33300	.86495	0.41500	.67341
0.17000	1.49770	0.25200	1.12097	0.33400	.86227	0.41600	.67139
0.17100	1.49200	0.25300	1.11726	0.33500	.85959	0.41700	.66937
0.17200	1.48634	0.25400	1.11356	0.33600	.85693	0.41800	.66736
0.17300	1.48071	0.25500	1.10988	0.33700	.85428	0.41900	.66535
0.17400	1.47512	0.25600	1.10621	0.33800	.85163	0.42000	.66335
0.17500	1.46956	0.25700	1.10256	0.33900	.84900	0.42100	.66136
0.17600	1.46403	0.25800	1.09892	0.34000	.84637	0.42200	.65937
0.17700	1.45854	0.25900	1.09531	0.34100	.84376	0.42300	.65739
0.17800	1.45308	0.26000	1.09170	0.34200	.84115	0.42400	.65542
0.17900	1.44765	0.26100	1.08812	0.34300	.83856	0.42500	.65345
0.18000	1.44226	0.26200	1.08454	0.34400	.83597	0.42600	.65149
0.18100	1.43690	0.26300	1.08099	0.34500	.83339	0.42700	.64954
0.18200	1.43157	0.26400	1.07745	0.34600	.83082	0.42800	.64759
0.18300	1.42627	0.26500	1.07392	0.34700	.82826	0.42900	.64564
0.18400	1.42100	0.26600	1.07041	0.34800	.82571	0.43000	.64371
0.18500	1.41576	0.26700	1.06692	0.34900	.82317	0.43100	.64178
0.18600	1.41055	0.26800	1.06344	0.35000	.82064	0.43200	.63985
0.18700	1.40537	0.26900	1.05997	0.35100	.81811	0.43300	.63794
0.18800	1.40022	0.27000	1.05652	0.35200	.81560	0.43400	.63603
0.18900	1.39510	0.27100	1.05309	0.35300	.81310	0.43500	.63412
0.19000	1.39001	0.27200	1.04966	0.35400	.81060	0.43600	.63222
0.19100	1.38495	0.27300	1.04626	0.35500	.80811	0.43700	.63033
0.19200	1.37992	0.27400	1.04286	0.35600	.80563	0.43800	.62844
0.19300	1.37491	0.27500	1.03949	0.35700	.80316	0.43900	.62656
0.19400	1.36993	0.27600	1.03612	0.35800	.80070	0.44000	.62468
0.19500	1.36498	0.27700	1.03277	0.35900	.79825	0.44100	.62281
0.19600	1.36006	0.27800	1.02943	0.36000	.79580	0.44200	.62095
0.19700	1.35516	0.27900	1.02611	0.36100	.79337	0.44300	.61909
0.19800	1.35029	0.28000	1.02280	0.36200	.79094	0.44400	.61724
0.19900	1.34545	0.28100	1.01951	0.36300	.78852	0.44500	.61539
0.20000	1.34063	0.28200	1.01623	0.36400	.78611	0.44600	.61355
0.20100	1.33584	0.28300	1.01296	0.36500	.78371	0.44700	.61172
0.20200	1.33108	0.28400	1.00970	0.36600	.78131	0.44800	.60989
0.20300	1.32634	0.28500	1.00646	0.36700	.77893	0.44900	.60806
0.20400	1.32162	0.28600	1.00323	0.36800	.77655	0.45000	.60625
0.20500	1.31693	0.28700	1.00002	0.36900	.77418	0.45100	.60443
0.20600	1.31227	0.28800	.99681	0.37000	.77182	0.45200	.60263
0.20700	1.30763	0.28900	.99362	0.37100	.76947	0.45300	.60083

$\frac{r}{\sqrt{4Tt}}$	$\int_{r/\sqrt{4Tt}}^{\infty} \frac{1}{u} e^{-u^2} du$	$\frac{r}{\sqrt{4Tt}}$	$\int_{r/\sqrt{4Tt}}^{\infty} \frac{1}{u} e^{-u^2} du$	$\frac{r}{\sqrt{4Tt}}$	$\int_{r/\sqrt{4Tt}}^{\infty} \frac{1}{u} e^{-u^2} du$	$\frac{r}{\sqrt{4Tt}}$	$\int_{r/\sqrt{4Tt}}^{\infty} \frac{1}{u} e^{-u^2} du$
0.45400	.59903	0.53600	.46897	0.61800	.36683	0.70000	.28604
0.45500	.59724	0.53700	.46757	0.61900	.36573	0.70100	.28517
0.45600	.59546	0.53800	.46618	0.62000	.36463	0.70200	.28430
0.45700	.59368	0.53900	.46479	0.62100	.36353	0.70300	.28343
0.45800	.59191	0.54000	.46340	0.62200	.36244	0.70400	.28256
0.45900	.59014	0.54100	.46202	0.62300	.36135	0.70500	.28170
0.46000	.58838	0.54200	.46064	0.62400	.36026	0.70600	.28084
0.46100	.58662	0.54300	.45927	0.62500	.35918	0.70700	.27998
0.46200	.58487	0.54400	.45790	0.62600	.35810	0.70800	.27912
0.46300	.58312	0.54500	.45654	0.62700	.35702	0.70900	.27827
0.46400	.58138	0.54600	.45517	0.62800	.35594	0.71000	.27742
0.46500	.57965	0.54700	.45382	0.62900	.35487	0.71100	.27657
0.46600	.57792	0.54800	.45246	0.63000	.35380	0.71200	.27572
0.46700	.57619	0.54900	.45111	0.63100	.35274	0.71300	.27487
0.46800	.57447	0.55000	.44977	0.63200	.35167	0.71400	.27403
0.46900	.57276	0.55100	.44843	0.63300	.35061	0.71500	.27319
0.47000	.57105	0.55200	.44709	0.63400	.34956	0.71600	.27235
0.47100	.56935	0.55300	.44575	0.63500	.34850	0.71700	.27152
0.47200	.56765	0.55400	.44442	0.63600	.34745	0.71800	.27069
0.47300	.56596	0.55500	.44310	0.63700	.34640	0.71900	.26985
0.47400	.56427	0.55600	.44178	0.63800	.34536	0.72000	.26903
0.47500	.56259	0.55700	.44046	0.63900	.34432	0.72100	.26820
0.47600	.56091	0.55800	.43914	0.64000	.34328	0.72200	.26738
0.47700	.55924	0.55900	.43783	0.64100	.34224	0.72300	.26656
0.47800	.55757	0.56000	.43653	0.64200	.34121	0.72400	.26574
0.47900	.55591	0.56100	.43522	0.64300	.34018	0.72500	.26492
0.48000	.55425	0.56200	.43392	0.64400	.33915	0.72600	.26411
0.48100	.55260	0.56300	.43263	0.64500	.33813	0.72700	.26329
0.48200	.55095	0.56400	.43134	0.64600	.33711	0.72800	.26248
0.48300	.54931	0.56500	.43005	0.64700	.33609	0.72900	.26168
0.48400	.54767	0.56600	.42876	0.64800	.33507	0.73000	.26087
0.48500	.54604	0.56700	.42748	0.64900	.33406	0.73100	.26007
0.48600	.54441	0.56800	.42621	0.65000	.33305	0.73200	.25927
0.48700	.54279	0.56900	.42493	0.65100	.33204	0.73300	.25847
0.48800	.54117	0.57000	.42366	0.65200	.33104	0.73400	.25767
0.48900	.53956	0.57100	.42240	0.65300	.33004	0.73500	.25688
0.49000	.53795	0.57200	.42114	0.65400	.32904	0.73600	.25609
0.49100	.53635	0.57300	.41988	0.65500	.32805	0.73700	.25530
0.49200	.53475	0.57400	.41862	0.65600	.32705	0.73800	.25451
0.49300	.53316	0.57500	.41737	0.65700	.32606	0.73900	.25373
0.49400	.53157	0.57600	.41612	0.65800	.32508	0.74000	.25294
0.49500	.52999	0.57700	.41488	0.65900	.32409	0.74100	.25216
0.49600	.52841	0.57800	.41364	0.66000	.32311	0.74200	.25139
0.49700	.52684	0.57900	.41240	0.66100	.32213	0.74300	.25061
0.49800	.52527	0.58000	.41117	0.66200	.32116	0.74400	.24984
0.49900	.52070	0.50100	.40994	0.66300	.32018	0.74500	.24906
0.50000	.52214	0.58200	.40871	0.66400	.31921	0.74600	.24830
0.50100	.52059	0.58300	.40749	0.66500	.31824	0.74700	.24753
0.50200	.51904	0.58400	.40627	0.66600	.31728	0.74800	.24676
0.50300	.51749	0.58500	.40505	0.66700	.31632	0.74900	.24600
0.50400	.51595	0.58600	.40384	0.66800	.31536	0.75000	.24524
0.50500	.51441	0.58700	.40263	0.66900	.31440	0.75100	.24448
0.50600	.51288	0.58800	.40143	0.67000	.31345	0.75200	.24372
0.50700	.51135	0.58900	.40023	0.67100	.31249	0.75300	.24297
0.50800	.50983	0.59000	.39903	0.67200	.31155	0.75400	.24222
0.50900	.50831	0.59100	.39783	0.67300	.31060	0.75500	.24147
0.51000	.50680	0.59200	.39664	0.67400	.30966	0.75600	.24072
0.51100	.50529	0.59300	.39545	0.67500	.30872	0.75700	.23997
0.51200	.50378	0.59400	.39427	0.67600	.30778	0.75800	.23923
0.51300	.50228	0.59500	.39309	0.67700	.30684	0.75900	.23849
0.51400	.50078	0.59600	.39191	0.67800	.30591	0.76000	.23775
0.51500	.49929	0.59700	.39073	0.67900	.30498	0.76100	.23701
0.51600	.49781	0.59800	.38956	0.68000	.30405	0.76200	.23628
0.51700	.49632	0.59900	.38840	0.68100	.30313	0.76300	.23554
0.51800	.49485	0.60000	.38723	0.68200	.30221	0.76400	.23481
0.51900	.49337	0.60100	.38607	0.68300	.30129	0.76500	.23408
0.52000	.49190	0.60200	.38491	0.68400	.30037	0.76600	.23336
0.52100	.49044	0.60300	.38376	0.68500	.29945	0.76700	.23263
0.52200	.48898	0.60400	.38261	0.68600	.29854	0.76800	.23191
0.52300	.48752	0.60500	.38146	0.68700	.29763	0.76900	.23119
0.52400	.48607	0.60600	.38031	0.68800	.29673	0.77000	.23047
0.52500	.48462	0.60700	.37917	0.68900	.29582	0.77100	.22975
0.52600	.48317	0.60800	.37803	0.69000	.29492	0.77200	.22904
0.52700	.48173	0.60900	.37690	0.69100	.29402	0.77300	.22832
0.52800	.48030	0.61000	.37577	0.69200	.29313	0.77400	.22761
0.52900	.47887	0.61100	.37464	0.69300	.29223	0.77500	.22690
0.53000	.47744	0.61200	.37351	0.69400	.29134	0.77600	.22620
0.53100	.47602	0.61300	.37239	0.69500	.29045	0.77700	.22549
0.53200	.47460	0.61400	.37127	0.69600	.28956	0.77800	.22479
0.53300	.47319	0.61500	.37016	0.69700	.28868	0.77900	.22409
0.53400	.47178	0.61600	.36905	0.69800	.28780	0.78000	.22339
0.53500	.47037	0.61700	.36794	0.69900	.28692	0.78100	.22269

$\frac{r}{\sqrt{4Tt}}$	$\int_{r/\sqrt{4Tt}}^{\infty} \frac{1}{u} e^{-u^2} du$	$\frac{r}{\sqrt{4Tt}}$	$\int_{r/\sqrt{4Tt}}^{\infty} \frac{1}{u} e^{-u^2} du$	$\frac{r}{\sqrt{4Tt}}$	$\int_{r/\sqrt{4Tt}}^{\infty} \frac{1}{u} e^{-u^2} du$	$\frac{r}{\sqrt{4Tt}}$	$\int_{r/\sqrt{4Tt}}^{\infty} \frac{1}{u} e^{-u^2} du$
0.78200	.22200	0.86400	.17128	0.94600	.13125	1.02800	.09981
0.78300	.22131	0.86500	.17073	0.94700	.13082	1.02900	.09948
0.78400	.22062	0.86600	.17018	0.94800	.13038	1.03000	.09914
0.78500	.21993	0.86700	.16964	0.94900	.12996	1.03100	.09880
0.78600	.21924	0.86800	.16910	0.95000	.12953	1.03200	.09847
0.78700	.21855	0.86900	.16855	0.95100	.12910	1.03300	.09814
0.78800	.21787	0.87000	.16801	0.95200	.12868	1.03400	.09780
0.78900	.21719	0.87100	.16748	0.95300	.12825	1.03500	.09747
0.79000	.21651	0.87200	.16694	0.95400	.12783	1.03600	.09714
0.79100	.21583	0.87300	.16640	0.95500	.12741	1.03700	.09681
0.79200	.21516	0.87400	.16587	0.95600	.12699	1.03800	.09648
0.79300	.21449	0.87500	.16534	0.95700	.12657	1.03900	.09616
0.79400	.21381	0.87600	.16481	0.95800	.12615	1.04000	.09583
0.79500	.21314	0.87700	.16428	0.95900	.12574	1.04100	.09550
0.79600	.21248	0.87800	.16375	0.96000	.12532	1.04200	.09518
0.79700	.21181	0.87900	.16322	0.96100	.12491	1.04300	.09486
0.79800	.21115	0.88000	.16270	0.96200	.12450	1.04400	.09453
0.79900	.21049	0.88100	.16218	0.96300	.12408	1.04500	.09421
0.80000	.20983	0.88200	.16166	0.96400	.12367	1.04600	.09389
0.80100	.20917	0.88300	.16114	0.96500	.12327	1.04700	.09357
0.80200	.20851	0.88400	.16062	0.96600	.12286	1.04800	.09325
0.80300	.20786	0.88500	.16010	0.96700	.12245	1.04900	.09293
0.80400	.20720	0.88600	.15958	0.96800	.12205	1.05000	.09262
0.80500	.20655	0.88700	.15907	0.96900	.12164	1.05100	.09230
0.80600	.20590	0.88800	.15856	0.97000	.12124	1.05200	.09199
0.80700	.20526	0.88900	.15805	0.97100	.12084	1.05300	.09167
0.80800	.20461	0.89000	.15754	0.97200	.12044	1.05400	.09136
0.80900	.20397	0.89100	.15703	0.97300	.12004	1.05500	.09105
0.81000	.20333	0.89200	.15652	0.97400	.11964	1.05600	.09074
0.81100	.20269	0.89300	.15602	0.97500	.11924	1.05700	.09043
0.81200	.20205	0.89400	.15551	0.97600	.11885	1.05800	.09012
0.81300	.20141	0.89500	.15501	0.97700	.11845	1.05900	.08981
0.81400	.20078	0.89600	.15451	0.97800	.11806	1.06000	.08950
0.81500	.20015	0.89700	.15401	0.97900	.11767	1.06100	.08920
0.81600	.19952	0.89800	.15351	0.98000	.11727	1.06200	.08889
0.81700	.19889	0.89900	.15302	0.98100	.11688	1.06300	.08859
0.81800	.19826	0.90000	.15252	0.98200	.11650	1.06400	.08828
0.81900	.19764	0.90100	.15203	0.98300	.11611	1.06500	.08798
0.82000	.19701	0.90200	.15154	0.98400	.11572	1.06600	.08768
0.82100	.19639	0.90300	.15104	0.98500	.11534	1.06700	.08738
0.82200	.19577	0.90400	.15056	0.98600	.11495	1.06800	.08708
0.82300	.19515	0.90500	.15007	0.98700	.11457	1.06900	.08678
0.82400	.19454	0.90600	.14958	0.98800	.11419	1.07000	.08648
0.82500	.19392	0.90700	.14910	0.98900	.11381	1.07100	.08619
0.82600	.19331	0.90800	.14861	0.99000	.11343	1.07200	.08589
0.82700	.19270	0.90900	.14813	0.99100	.11305	1.07300	.08559
0.82800	.19209	0.91000	.14765	0.99200	.11267	1.07400	.08530
0.82900	.19148	0.91100	.14717	0.99300	.11229	1.07500	.08501
0.83000	.19088	0.91200	.14669	0.99400	.11192	1.07600	.08471
0.83100	.19027	0.91300	.14622	0.99500	.11155	1.07700	.08442
0.83200	.18967	0.91400	.14574	0.99600	.11117	1.07800	.08413
0.83300	.18907	0.91500	.14527	0.99700	.11080	1.07900	.08384
0.83400	.18847	0.91600	.14479	0.99800	.11043	1.08000	.08355
0.83500	.18787	0.91700	.14432	0.99900	.11006	1.08100	.08327
0.83600	.18728	0.91800	.14385	1.00000	.10969	1.08200	.08298
0.83700	.18668	0.91900	.14339	1.00100	.10932	1.08300	.08269
0.83800	.18609	0.92000	.14292	1.00200	.10896	1.08400	.08241
0.83900	.18550	0.92100	.14245	1.00300	.10859	1.08500	.08212
0.84000	.18491	0.92200	.14199	1.00400	.10823	1.08600	.08184
0.84100	.18432	0.92300	.14153	1.00500	.10787	1.08700	.08156
0.84200	.18374	0.92400	.14106	1.00600	.10750	1.08800	.08127
0.84300	.18316	0.92500	.14060	1.00700	.10714	1.08900	.08099
0.84400	.18257	0.92600	.14015	1.00800	.10678	1.09000	.08071
0.84500	.18199	0.92700	.13969	1.00900	.10643	1.09100	.08043
0.84600	.18141	0.92800	.13923	1.01000	.10607	1.09200	.08016
0.84700	.18084	0.92900	.13878	1.01100	.10571	1.09300	.07988
0.84800	.18026	0.93000	.13832	1.01200	.10536	1.09400	.07960
0.84900	.17969	0.93100	.13787	1.01300	.10500	1.09500	.07933
0.85000	.17912	0.93200	.13742	1.01400	.10465	1.09600	.07905
0.85100	.17855	0.93300	.13697	1.01500	.10430	1.09700	.07878
0.85200	.17798	0.93400	.13652	1.01600	.10395	1.09800	.07850
0.85300	.17741	0.93500	.13608	1.01700	.10359	1.09900	.07823
0.85400	.17684	0.93600	.13563	1.01800	.10325	1.10000	.07796
0.85500	.17628	0.93700	.13519	1.01900	.10290	1.10100	.07769
0.85600	.17572	0.93800	.13474	1.02000	.10255	1.10200	.07742
0.85700	.17516	0.93900	.13430	1.02100	.10221	1.10300	.07715
0.85800	.17460	0.94000	.13386	1.02200	.10186	1.10400	.07688
0.85900	.17404	0.94100	.13342	1.02300	.10152	1.10500	.07662
0.86000	.17349	0.94200	.13298	1.02400	.10117	1.10600	.07635
0.86100	.17293	0.94300	.13255	1.02500	.10083	1.10700	.07608
0.86200	.17238	0.94400	.13211	1.02600	.10049	1.10800	.07582
0.86300	.17183	0.94500	.13168	1.02700	.10015	1.10900	.07555

$\frac{r}{\sqrt{4Tt}}$	$\int_{r/\sqrt{4Tt}}^{\infty} \frac{1}{u} e^{-u^2} du$	$\frac{r}{\sqrt{4Tt}}$	$\int_{r/\sqrt{4Tt}}^{\infty} \frac{1}{u} e^{-u^2} du$	$\frac{r}{\sqrt{4Tt}}$	$\int_{r/\sqrt{4Tt}}^{\infty} \frac{1}{u} e^{-u^2} du$	$\frac{r}{\sqrt{4Tt}}$	$\int_{r/\sqrt{4Tt}}^{\infty} \frac{1}{u} e^{-u^2} du$
1.11000	.07529	1.19200	.05630	1.27400	.04173	1.35600	.03063
1.11100	.07503	1.19300	.05610	1.27500	.04157	1.35700	.03051
1.11200	.07477	1.19400	.05590	1.27600	.04142	1.35800	.03040
1.11300	.07451	1.19500	.05570	1.27700	.04126	1.35900	.03028
1.11400	.07425	1.19600	.05550	1.27800	.04111	1.36000	.03016
1.11500	.07399	1.19700	.05530	1.27900	.04096	1.36100	.03005
1.11600	.07373	1.19800	.05510	1.28000	.04081	1.36200	.02993
1.11700	.07347	1.19900	.05490	1.28100	.04065	1.36300	.02982
1.11800	.07322	1.20000	.05470	1.28200	.04050	1.36400	.02970
1.11900	.07296	1.20100	.05451	1.28300	.04035	1.36500	.02959
1.12000	.07270	1.20200	.05431	1.28400	.04020	1.36600	.02948
1.12100	.07245	1.20300	.05411	1.28500	.04005	1.36700	.02936
1.12200	.07220	1.20400	.05392	1.28600	.03990	1.36800	.02925
1.12300	.07194	1.20500	.05373	1.28700	.03975	1.36900	.02914
1.12400	.07169	1.20600	.05353	1.28800	.03961	1.37000	.02903
1.12500	.07144	1.20700	.05334	1.28900	.03946	1.37100	.02892
1.12600	.07119	1.20800	.05315	1.29000	.03931	1.37200	.02880
1.12700	.07094	1.20900	.05295	1.29100	.03917	1.37300	.02869
1.12800	.07069	1.21000	.05276	1.29200	.03902	1.37400	.02858
1.12900	.07044	1.21100	.05257	1.29300	.03887	1.37500	.02847
1.13000	.07020	1.21200	.05238	1.29400	.03873	1.37600	.02836
1.13100	.06995	1.21300	.05219	1.29500	.03858	1.37700	.02825
1.13200	.06970	1.21400	.05200	1.29600	.03844	1.37800	.02815
1.13300	.06946	1.21500	.05181	1.29700	.03830	1.37900	.02804
1.13400	.06922	1.21600	.05163	1.29800	.03815	1.38000	.02793
1.13500	.06897	1.21700	.05144	1.29900	.03801	1.38100	.02782
1.13600	.06873	1.21800	.05125	1.30000	.03787	1.38200	.02771
1.13700	.06849	1.21900	.05107	1.30100	.03773	1.38300	.02761
1.13800	.06825	1.22000	.05088	1.30200	.03759	1.38400	.02750
1.13900	.06801	1.22100	.05070	1.30300	.03745	1.38500	.02739
1.14000	.06777	1.22200	.05051	1.30400	.03730	1.38600	.02729
1.14100	.06753	1.22300	.05033	1.30500	.03717	1.38700	.02718
1.14200	.06729	1.22400	.05015	1.30600	.03703	1.38800	.02708
1.14300	.06705	1.22500	.04996	1.30700	.03689	1.38900	.02697
1.14400	.06682	1.22600	.04978	1.30800	.03675	1.39000	.02687
1.14500	.06658	1.22700	.04960	1.30900	.03661	1.39100	.02676
1.14600	.06635	1.22800	.04942	1.31000	.03647	1.39200	.02666
1.14700	.06611	1.22900	.04924	1.31100	.03634	1.39300	.02656
1.14800	.06588	1.23000	.04906	1.31200	.03620	1.39400	.02646
1.14900	.06565	1.23100	.04888	1.31300	.03606	1.39500	.02635
1.15000	.06541	1.23200	.04870	1.31400	.03593	1.39600	.02625
1.15100	.06518	1.23300	.04853	1.31500	.03579	1.39700	.02615
1.15200	.06495	1.23400	.04835	1.31600	.03566	1.39800	.02605
1.15300	.06472	1.23500	.04817	1.31700	.03552	1.39900	.02595
1.15400	.06449	1.23600	.04800	1.31800	.03539	1.40000	.02584
1.15500	.06426	1.23700	.04782	1.31900	.03526	1.40100	.02574
1.15600	.06404	1.23800	.04765	1.32000	.03512	1.40200	.02564
1.15700	.06381	1.23900	.04747	1.32100	.03499	1.40300	.02554
1.15800	.06358	1.24000	.04730	1.32200	.03486	1.40400	.02545
1.15900	.06336	1.24100	.04713	1.32300	.03473	1.40500	.02535
1.16000	.06313	1.24200	.04695	1.32400	.03460	1.40600	.02525
1.16100	.06291	1.24300	.04678	1.32500	.03447	1.40700	.02515
1.16200	.06268	1.24400	.04661	1.32600	.03434	1.40800	.02505
1.16300	.06246	1.24500	.04644	1.32700	.03421	1.40900	.02495
1.16400	.06224	1.24600	.04627	1.32800	.03408	1.41000	.02486
1.16500	.06202	1.24700	.04610	1.32900	.03395	1.41100	.02476
1.16600	.06180	1.24800	.04593	1.33000	.03382	1.41200	.02466
1.16700	.06158	1.24900	.04576	1.33100	.03360	1.41300	.02457
1.16800	.06136	1.25000	.04559	1.33200	.03356	1.41400	.02447
1.16900	.06114	1.25100	.04543	1.33300	.03344	1.41500	.02437
1.17000	.06092	1.25200	.04526	1.33400	.03331	1.41600	.02428
1.17100	.06071	1.25300	.04509	1.33500	.03318	1.41700	.02418
1.17200	.06049	1.25400	.04493	1.33600	.03306	1.41800	.02409
1.17300	.06027	1.25500	.04476	1.33700	.03293	1.41900	.02400
1.17400	.06006	1.25600	.04460	1.33800	.03281	1.42000	.02390
1.17500	.05984	1.25700	.04443	1.33900	.03268	1.42100	.02381
1.17600	.05963	1.25800	.04427	1.34000	.03256	1.42200	.02372
1.17700	.05942	1.25900	.04411	1.34100	.03244	1.42300	.02362
1.17800	.05921	1.26000	.04394	1.34200	.03231	1.42400	.02353
1.17900	.05899	1.26100	.04378	1.34300	.03219	1.42500	.02344
1.18000	.05878	1.26200	.04362	1.34400	.03207	1.42600	.02335
1.18100	.05857	1.26300	.04346	1.34500	.03194	1.42700	.02325
1.18200	.05836	1.26400	.04330	1.34600	.03182	1.42800	.02316
1.18300	.05815	1.26500	.04314	1.34700	.03170	1.42900	.02307
1.18400	.05795	1.26600	.04298	1.34800	.03158	1.43000	.02298
1.18500	.05774	1.26700	.04282	1.34900	.03146	1.43100	.02289
1.18600	.05753	1.26800	.04266	1.35000	.03134	1.43200	.02280
1.18700	.05733	1.26900	.04251	1.35100	.03122	1.43300	.02271
1.18800	.05712	1.27000	.04235	1.35200	.03110	1.43400	.02262
1.18900	.05692	1.27100	.04219	1.35300	.03098	1.43500	.02253
1.19000	.05671	1.27200	.04204	1.35400	.03087	1.43600	.02244
1.19100	.05651	1.27300	.04188	1.35500	.03075	1.43700	.02236

$\frac{r}{\sqrt{4Tt}}$	$\int_{r/\sqrt{4Tt}}^{\infty} \frac{1}{u} e^{-u^2} du$	$\frac{r}{\sqrt{4Tt}}$	$\int_{r/\sqrt{4Tt}}^{\infty} \frac{1}{u} e^{-u^2} du$	$\frac{r}{\sqrt{4Tt}}$	$\int_{r/\sqrt{4Tt}}^{\infty} \frac{1}{u} e^{-u^2} du$	$\frac{r}{\sqrt{4Tt}}$	$\int_{r/\sqrt{4Tt}}^{\infty} \frac{1}{u} e^{-u^2} du$
1.43800	.02227	1.52000	.01603	1.60200	.01142	1.68400	.00805
1.43900	.02218	1.52100	.01596	1.60300	.01137	1.68500	.00801
1.44000	.02209	1.52200	.01590	1.60400	.01132	1.68600	.00798
1.44100	.02200	1.52300	.01583	1.60500	.01127	1.68700	.00794
1.44200	.02192	1.52400	.01577	1.60600	.01123	1.68800	.00791
1.44300	.02183	1.52500	.01570	1.60700	.01118	1.68900	.00787
1.44400	.02175	1.52600	.01564	1.60800	.01113	1.69000	.00784
1.44500	.02166	1.52700	.01557	1.60900	.01109	1.69100	.00781
1.44600	.02157	1.52800	.01551	1.61000	.01104	1.69200	.00777
1.44700	.02149	1.52900	.01545	1.61100	.01099	1.69300	.00774
1.44800	.02140	1.53000	.01539	1.61200	.01095	1.69400	.00771
1.44900	.02132	1.53100	.01532	1.61300	.01090	1.69500	.00767
1.45000	.02123	1.53200	.01526	1.61400	.01085	1.69600	.00764
1.45100	.02115	1.53300	.01520	1.61500	.01081	1.69700	.00761
1.45200	.02107	1.53400	.01514	1.61600	.01076	1.69800	.00757
1.45300	.02098	1.53500	.01507	1.61700	.01072	1.69900	.00754
1.45400	.02090	1.53600	.01501	1.61800	.01067	1.70000	.00751
1.45500	.02082	1.53700	.01495	1.61900	.01063	1.70100	.00747
1.45600	.02073	1.53800	.01489	1.62000	.01058	1.70200	.00744
1.45700	.02065	1.53900	.01483	1.62100	.01054	1.70300	.00741
1.45800	.02057	1.54000	.01477	1.62200	.01049	1.70400	.00738
1.45900	.02049	1.54100	.01471	1.62300	.01045	1.70500	.00735
1.46000	.02041	1.54200	.01465	1.62400	.01040	1.70600	.00731
1.46100	.02033	1.54300	.01459	1.62500	.01036	1.70700	.00728
1.46200	.02024	1.54400	.01453	1.62600	.01032	1.70800	.00725
1.46300	.02016	1.54500	.01447	1.62700	.01027	1.70900	.00722
1.46400	.02008	1.54600	.01441	1.62800	.01023	1.71000	.00719
1.46500	.02000	1.54700	.01435	1.62900	.01019	1.71100	.00716
1.46600	.01992	1.54800	.01429	1.63000	.01014	1.71200	.00712
1.46700	.01985	1.54900	.01423	1.63100	.01010	1.71300	.00709
1.46800	.01977	1.55000	.01417	1.63200	.01006	1.71400	.00706
1.46900	.01969	1.55100	.01411	1.63300	.01001	1.71500	.00703
1.47000	.01961	1.55200	.01406	1.63400	.00997	1.71600	.00700
1.47100	.01953	1.55300	.01400	1.63500	.00993	1.71700	.00697
1.47200	.01945	1.55400	.01394	1.63600	.00989	1.71800	.00694
1.47300	.01937	1.55500	.01388	1.63700	.00985	1.71900	.00691
1.47400	.01930	1.55600	.01383	1.63800	.00980	1.72000	.00688
1.47500	.01922	1.55700	.01377	1.63900	.00976	1.72100	.00685
1.47600	.01914	1.55800	.01371	1.64000	.00972	1.72200	.00682
1.47700	.01907	1.55900	.01366	1.64100	.00968	1.72300	.00679
1.47800	.01899	1.56000	.01360	1.64200	.00964	1.72400	.00676
1.47900	.01891	1.56100	.01354	1.64300	.00960	1.72500	.00673
1.48000	.01884	1.56200	.01349	1.64400	.00956	1.72600	.00670
1.48100	.01876	1.56300	.01343	1.64500	.00952	1.72700	.00667
1.48200	.01869	1.56400	.01338	1.64600	.00948	1.72800	.00664
1.48300	.01861	1.56500	.01332	1.64700	.00944	1.72900	.00661
1.48400	.01854	1.56600	.01327	1.64800	.00939	1.73000	.00658
1.48500	.01846	1.56700	.01321	1.64900	.00935	1.73100	.00655
1.48600	.01839	1.56800	.01316	1.65000	.00932	1.73200	.00653
1.48700	.01832	1.56900	.01310	1.65100	.00928	1.73300	.00650
1.48800	.01824	1.57000	.01305	1.65200	.00924	1.73400	.00647
1.48900	.01817	1.57100	.01299	1.65300	.00920	1.73500	.00644
1.49000	.01810	1.57200	.01294	1.65400	.00916	1.73600	.00641
1.49100	.01802	1.57300	.01289	1.65500	.00912	1.73700	.00638
1.49200	.01795	1.57400	.01283	1.65600	.00908	1.73800	.00636
1.49300	.01788	1.57500	.01278	1.65700	.00904	1.73900	.00633
1.49400	.01781	1.57600	.01273	1.65800	.00900	1.74000	.00630
1.49500	.01774	1.57700	.01267	1.65900	.00896	1.74100	.00627
1.49600	.01766	1.57800	.01262	1.66000	.00892	1.74200	.00624
1.49700	.01759	1.57900	.01257	1.66100	.00889	1.74300	.00622
1.49800	.01752	1.58000	.01252	1.66200	.00885	1.74400	.00619
1.49900	.01745	1.58100	.01246	1.66300	.00881	1.74500	.00616
1.50000	.01738	1.58200	.01241	1.66400	.00877	1.74600	.00613
1.50100	.01731	1.58300	.01236	1.66500	.00873	1.74700	.00611
1.50200	.01724	1.58400	.01231	1.66600	.00870	1.74800	.00608
1.50300	.01717	1.58500	.01226	1.66700	.00866	1.74900	.00605
1.50400	.01710	1.58600	.01221	1.66800	.00862	1.75000	.00603
1.50500	.01703	1.58700	.01216	1.66900	.00859	1.75100	.00600
1.50600	.01696	1.58800	.01211	1.67000	.00855	1.75200	.00597
1.50700	.01690	1.58900	.01206	1.67100	.00851	1.75300	.00595
1.50800	.01683	1.59000	.01200	1.67200	.00848	1.75400	.00592
1.50900	.01676	1.59100	.01195	1.67300	.00844	1.75500	.00589
1.51000	.01669	1.59200	.01190	1.67400	.00840	1.75600	.00587
1.51100	.01662	1.59300	.01186	1.67500	.00837	1.75700	.00584
1.51200	.01656	1.59400	.01181	1.67600	.00833	1.75800	.00582
1.51300	.01649	1.59500	.01176	1.67700	.00829	1.75900	.00579
1.51400	.01642	1.59600	.01171	1.67800	.00826	1.76000	.00576
1.51500	.01636	1.59700	.01166	1.67900	.00822	1.76100	.00574
1.51600	.01629	1.59800	.01161	1.68000	.00819	1.76200	.00571
1.51700	.01622	1.59900	.01156	1.68100	.00815	1.76300	.00569
1.51800	.01616	1.60000	.01151	1.68200	.00812	1.76400	.00566
1.51900	.01609	1.60100	.01146	1.68300	.00808	1.76500	.00564

$\frac{r}{\sqrt{4Tt}}$	$\int_{r/\sqrt{4Tt}}^{\infty} \frac{1}{u} e^{-u^2} du$	$\frac{r}{\sqrt{4Tt}}$	$\int_{r/\sqrt{4Tt}}^{\infty} \frac{1}{u} e^{-u^2} du$	$\frac{r}{\sqrt{4Tt}}$	$\int_{r/\sqrt{4Tt}}^{\infty} \frac{1}{u} e^{-u^2} du$	$\frac{r}{\sqrt{4Tt}}$	$\int_{r/\sqrt{4Tt}}^{\infty} \frac{1}{u} e^{-u^2} du$
1.76600	.00561	1.84800	.00387	1.93000	.00264	2.01200	.00178
1.76700	.00559	1.84900	.00385	1.93100	.00263	2.01300	.00177
1.76800	.00556	1.85000	.00384	1.93200	.00262	2.01400	.00177
1.76900	.00554	1.85100	.00382	1.93300	.00260	2.01500	.00176
1.77000	.00551	1.85200	.00380	1.93400	.00259	2.01600	.00175
1.77100	.00549	1.85300	.00378	1.93500	.00258	2.01700	.00174
1.77200	.00546	1.85400	.00377	1.93600	.00257	2.01800	.00173
1.77300	.00544	1.85500	.00375	1.93700	.00256	2.01900	.00172
1.77400	.00542	1.85600	.00373	1.93800	.00254	2.02000	.00171
1.77500	.00539	1.85700	.00371	1.93900	.00253	2.02100	.00171
1.77600	.00537	1.85800	.00370	1.94000	.00252	2.02200	.00170
1.77700	.00534	1.85900	.00368	1.94100	.00251	2.02300	.00169
1.77800	.00532	1.86000	.00366	1.94200	.00250	2.02400	.00168
1.77900	.00530	1.86100	.00365	1.94300	.00248	2.02500	.00167
1.78000	.00527	1.86200	.00363	1.94400	.00247	2.02600	.00166
1.78100	.00525	1.86300	.00361	1.94500	.00246	2.02700	.00166
1.78200	.00522	1.86400	.00360	1.94600	.00245	2.02800	.00165
1.78300	.00520	1.86500	.00358	1.94700	.00244	2.02900	.00164
1.78400	.00518	1.86600	.00356	1.94800	.00243	2.03000	.00163
1.78500	.00515	1.86700	.00355	1.94900	.00241	2.03100	.00162
1.78600	.00513	1.86800	.00353	1.95000	.00240	2.03200	.00162
1.78700	.00511	1.86900	.00351	1.95100	.00239	2.03300	.00161
1.78800	.00509	1.87000	.00350	1.95200	.00238	2.03400	.00160
1.78900	.00506	1.87100	.00348	1.95300	.00237	2.03500	.00159
1.79000	.00504	1.87200	.00347	1.95400	.00236	2.03600	.00159
1.79100	.00502	1.87300	.00345	1.95500	.00235	2.03700	.00158
1.79200	.00500	1.87400	.00343	1.95600	.00233	2.03800	.00157
1.79300	.00497	1.87500	.00342	1.95700	.00232	2.03900	.00156
1.79400	.00495	1.87600	.00340	1.95800	.00231	2.04000	.00155
1.79500	.00493	1.87700	.00339	1.95900	.00230	2.04100	.00155
1.79600	.00491	1.87800	.00337	1.96000	.00229	2.04200	.00154
1.79700	.00488	1.87900	.00336	1.96100	.00228	2.04300	.00153
1.79800	.00486	1.88000	.00334	1.96200	.00227	2.04400	.00152
1.79900	.00484	1.88100	.00332	1.96300	.00226	2.04500	.00152
1.80000	.00482	1.88200	.00331	1.96400	.00225	2.04600	.00151
1.80100	.00480	1.88300	.00329	1.96500	.00224	2.04700	.00150
1.80200	.00477	1.88400	.00328	1.96600	.00223	2.04800	.00149
1.80300	.00475	1.88500	.00326	1.96700	.00222	2.04900	.00149
1.80400	.00473	1.88600	.00325	1.96800	.00220	2.05000	.00148
1.80500	.00471	1.88700	.00323	1.96900	.00219	2.05100	.00147
1.80600	.00469	1.88800	.00322	1.97000	.00218	2.05200	.00147
1.80700	.00467	1.88900	.00320	1.97100	.00217	2.05300	.00146
1.80800	.00465	1.89000	.00319	1.97200	.00216	2.05400	.00145
1.80900	.00463	1.89100	.00317	1.97300	.00215	2.05500	.00144
1.81000	.00461	1.89200	.00316	1.97400	.00214	2.05600	.00144
1.81100	.00458	1.89300	.00314	1.97500	.00213	2.05700	.00143
1.81200	.00456	1.89400	.00313	1.97600	.00212	2.05800	.00142
1.81300	.00454	1.89500	.00311	1.97700	.00211	2.05900	.00142
1.81400	.00452	1.89600	.00310	1.97800	.00210	2.06000	.00141
1.81500	.00450	1.89700	.00309	1.97900	.00209	2.06100	.00140
1.81600	.00448	1.89800	.00307	1.98000	.00208	2.06200	.00139
1.81700	.00446	1.89900	.00306	1.98100	.00207	2.06300	.00139
1.81800	.00444	1.90000	.00304	1.98200	.00206	2.06400	.00138
1.81900	.00442	1.90100	.00303	1.98300	.00205	2.06500	.00137
1.82000	.00440	1.90200	.00301	1.98400	.00204	2.06600	.00137
1.82100	.00438	1.90300	.00300	1.98500	.00203	2.06700	.00136
1.82200	.00436	1.90400	.00299	1.98600	.00202	2.06800	.00135
1.82300	.00434	1.90500	.00297	1.98700	.00201	2.06900	.00135
1.82400	.00432	1.90600	.00296	1.98800	.00200	2.07000	.00134
1.82500	.00430	1.90700	.00294	1.98900	.00199	2.07100	.00133
1.82600	.00428	1.90800	.00293	1.99000	.00198	2.07200	.00133
1.82700	.00426	1.90900	.00292	1.99100	.00197	2.07300	.00132
1.82800	.00424	1.91000	.00290	1.99200	.00196	2.07400	.00131
1.82900	.00422	1.91100	.00289	1.99300	.00195	2.07500	.00131
1.83000	.00420	1.91200	.00288	1.99400	.00195	2.07600	.00130
1.83100	.00419	1.91300	.00286	1.99500	.00194	2.07700	.00129
1.83200	.00417	1.91400	.00285	1.99600	.00193	2.07800	.00129
1.83300	.00415	1.91500	.00284	1.99700	.00192	2.07900	.00128
1.83400	.00413	1.91600	.00282	1.99800	.00191	2.08000	.00128
1.83500	.00411	1.91700	.00281	1.99900	.00190	2.08100	.00127
1.83600	.00409	1.91800	.00280	2.00000	.00189	2.08200	.00126
1.83700	.00407	1.91900	.00278	2.00100	.00188	2.08300	.00126
1.83800	.00405	1.92000	.00277	2.00200	.00187	2.08400	.00125
1.83900	.00404	1.92100	.00276	2.00300	.00186	2.08500	.00124
1.84000	.00402	1.92200	.00274	2.00400	.00185	2.08600	.00124
1.84100	.00400	1.92300	.00273	2.00500	.00184	2.08700	.00123
1.84200	.00398	1.92400	.00272	2.00600	.00184	2.08800	.00123
1.84300	.00396	1.92500	.00271	2.00700	.00183	2.08900	.00122
1.84400	.00394	1.92600	.00269	2.00800	.00182	2.09000	.00121
1.84500	.00393	1.92700	.00268	2.00900	.00181	2.09100	.00121
1.84600	.00391	1.92800	.00267	2.01000	.00180	2.09200	.00120
1.84700	.00389	1.92900	.00265	2.01100	.00179	2.09300	.00120

$\frac{r}{\sqrt{4Tt}}$	$\int_{r/\sqrt{4Tt}}^{\infty} \frac{1}{u} e^{-u^2} du$	$\frac{r}{\sqrt{4Tt}}$	$\int_{r/\sqrt{4Tt}}^{\infty} \frac{1}{u} e^{-u^2} du$	$\frac{r}{\sqrt{4Tt}}$	$\int_{r/\sqrt{4Tt}}^{\infty} \frac{1}{u} e^{-u^2} du$	$\frac{r}{\sqrt{4Tt}}$	$\int_{r/\sqrt{4Tt}}^{\infty} \frac{1}{u} e^{-u^2} du$
2.09400	.00119	2.17600	.00078	2.25800	.00051	2.34000	.00033
2.09500	.00118	2.17700	.00078	2.25900	.00051	2.34100	.00033
2.09600	.00118	2.17800	.00078	2.26000	.00051	2.34200	.00033
2.09700	.00117	2.17900	.00077	2.26100	.00050	2.34300	.00032
2.09800	.00117	2.18000	.00077	2.26200	.00050	2.34400	.00032
2.09900	.00116	2.18100	.00076	2.26300	.00050	2.34500	.00032
2.10000	.00115	2.18200	.00076	2.26400	.00050	2.34600	.00032
2.10100	.00115	2.18300	.00076	2.26500	.00049	2.34700	.00032
2.10200	.00114	2.18400	.00075	2.26600	.00049	2.34800	.00032
2.10300	.00114	2.18500	.00075	2.26700	.00049	2.34900	.00031
2.10400	.00113	2.18600	.00074	2.26800	.00048	2.35000	.00031
2.10500	.00113	2.18700	.00074	2.26900	.00048	2.35100	.00031
2.10600	.00112	2.18800	.00074	2.27000	.00048	2.35200	.00031
2.10700	.00111	2.18900	.00073	2.27100	.00048	2.35300	.00031
2.10800	.00111	2.19000	.00073	2.27200	.00047	2.35400	.00031
2.10900	.00110	2.19100	.00073	2.27300	.00047	2.35500	.00030
2.11000	.00110	2.19200	.00072	2.27400	.00047	2.35600	.00030
2.11100	.00109	2.19300	.00072	2.27500	.00047	2.35700	.00030
2.11200	.00109	2.19400	.00071	2.27600	.00046	2.35800	.00030
2.11300	.00108	2.19500	.00071	2.27700	.00046	2.35900	.00030
2.11400	.00108	2.19600	.00071	2.27800	.00046	2.36000	.00030
2.11500	.00107	2.19700	.00070	2.27900	.00046	2.36100	.00029
2.11600	.00106	2.19800	.00070	2.28000	.00045	2.36200	.00029
2.11700	.00106	2.19900	.00070	2.28100	.00045	2.36300	.00029
2.11800	.00105	2.20000	.00069	2.28200	.00045	2.36400	.00029
2.11900	.00105	2.20100	.00069	2.28300	.00045	2.36500	.00029
2.12000	.00104	2.20200	.00069	2.28400	.00045	2.36600	.00029
2.12100	.00104	2.20300	.00068	2.28500	.00044	2.36700	.00028
2.12200	.00103	2.20400	.00068	2.28600	.00044	2.36800	.00028
2.12300	.00103	2.20500	.00068	2.28700	.00044	2.36900	.00028
2.12400	.00102	2.20600	.00067	2.28800	.00044	2.37000	.00028
2.12500	.00102	2.20700	.00067	2.28900	.00043	2.37100	.00028
2.12600	.00101	2.20800	.00066	2.29000	.00043	2.37200	.00028
2.12700	.00101	2.20900	.00066	2.29100	.00043	2.37300	.00028
2.12800	.00100	2.21000	.00066	2.29200	.00043	2.37400	.00027
2.12900	.00100	2.21100	.00065	2.29300	.00042	2.37500	.00027
2.13000	.00099	2.21200	.00065	2.29400	.00042	2.37600	.00027
2.13100	.00099	2.21300	.00065	2.29500	.00042	2.37700	.00027
2.13200	.00098	2.21400	.00064	2.29600	.00042	2.37800	.00027
2.13300	.00098	2.21500	.00064	2.29700	.00042	2.37900	.00027
2.13400	.00097	2.21600	.00064	2.29800	.00041	2.38000	.00026
2.13500	.00097	2.21700	.00063	2.29900	.00041	2.38100	.00026
2.13600	.00096	2.21800	.00063	2.30000	.00041	2.38200	.00026
2.13700	.00096	2.21900	.00063	2.30100	.00041	2.38300	.00026
2.13800	.00095	2.22000	.00062	2.30200	.00040	2.38400	.00026
2.13900	.00095	2.22100	.00062	2.30300	.00040	2.38500	.00026
2.14000	.00094	2.22200	.00062	2.30400	.00040	2.38600	.00026
2.14100	.00094	2.22300	.00061	2.30500	.00040	2.38700	.00025
2.14200	.00093	2.22400	.00061	2.30600	.00040	2.38800	.00025
2.14300	.00093	2.22500	.00061	2.30700	.00039	2.38900	.00025
2.14400	.00092	2.22600	.00061	2.30800	.00039	2.39000	.00025
2.14500	.00092	2.22700	.00060	2.30900	.00039	2.39100	.00025
2.14600	.00091	2.22800	.00060	2.31000	.00039	2.39200	.00025
2.14700	.00091	2.22900	.00060	2.31100	.00039	2.39300	.00025
2.14800	.00091	2.23000	.00059	2.31200	.00038	2.39400	.00025
2.14900	.00090	2.23100	.00059	2.31300	.00038	2.39500	.00024
2.15000	.00090	2.23200	.00059	2.31400	.00038	2.39600	.00024
2.15100	.00089	2.23300	.00058	2.31500	.00038	2.39700	.00024
2.15200	.00089	2.23400	.00058	2.31600	.00038	2.39800	.00024
2.15300	.00088	2.23500	.00058	2.31700	.00037	2.39900	.00024
2.15400	.00088	2.23600	.00057	2.31800	.00037	2.40000	.00024
2.15500	.00087	2.23700	.00057	2.31900	.00037	2.40100	.00024
2.15600	.00087	2.23800	.00057	2.32000	.00037	2.40200	.00023
2.15700	.00086	2.23900	.00057	2.32100	.00037	2.40300	.00023
2.15800	.00086	2.24000	.00056	2.32200	.00036	2.40400	.00023
2.15900	.00086	2.24100	.00056	2.32300	.00036	2.40500	.00023
2.16000	.00085	2.24200	.00056	2.32400	.00036	2.40600	.00023
2.16100	.00085	2.24300	.00055	2.32500	.00036	2.40700	.00023
2.16200	.00084	2.24400	.00055	2.32600	.00036	2.40800	.00023
2.16300	.00084	2.24500	.00055	2.32700	.00035	2.40900	.00023
2.16400	.00083	2.24600	.00054	2.32800	.00035	2.41000	.00022
2.16500	.00083	2.24700	.00054	2.32900	.00035	2.41100	.00022
2.16600	.00083	2.24800	.00054	2.33000	.00035	2.41200	.00022
2.16700	.00082	2.24900	.00054	2.33100	.00035	2.41300	.00022
2.16800	.00082	2.25000	.00053	2.33200	.00034	2.41400	.00022
2.16900	.00081	2.25100	.00053	2.33300	.00034	2.41500	.00022
2.17000	.00081	2.25200	.00053	2.33400	.00034	2.41600	.00022
2.17100	.00080	2.25300	.00052	2.33500	.00034	2.41700	.00022
2.17200	.00080	2.25400	.00052	2.33600	.00034	2.41800	.00021
2.17300	.00080	2.25500	.00052	2.33700	.00033	2.41900	.00021
2.17400	.00079	2.25600	.00052	2.33800	.00033	2.42000	.00021
2.17500	.00079	2.25700	.00051	2.33900	.00033	2.42100	.00021

$\frac{r}{\sqrt{4Tt}}$	$\int_{r/\sqrt{4Tt}}^{\infty} \frac{1}{u} e^{-u^2} du$	$\frac{r}{\sqrt{4Tt}}$	$\int_{r/\sqrt{4Tt}}^{\infty} \frac{1}{u} e^{-u^2} du$	$\frac{r}{\sqrt{4Tt}}$	$\int_{r/\sqrt{4Tt}}^{\infty} \frac{1}{u} e^{-u^2} du$	$\frac{r}{\sqrt{4Tt}}$	$\int_{r/\sqrt{4Tt}}^{\infty} \frac{1}{u} e^{-u^2} du$
2.42200	.00021	2.50400	.00013	2.58600	.00008	2.66800	.00005
2.42300	.00021	2.50500	.00013	2.58700	.00008	2.66900	.00005
2.42400	.00021	2.50600	.00013	2.58800	.00008	2.67000	.00005
2.42500	.00021	2.50700	.00013	2.58900	.00008	2.67100	.00005
2.42600	.00021	2.50800	.00013	2.59000	.00008	2.67200	.00005
2.42700	.00020	2.50900	.00013	2.59100	.00008	2.67300	.00005
2.42800	.00020	2.51000	.00013	2.59200	.00008	2.67400	.00005
2.42900	.00020	2.51100	.00013	2.59300	.00008	2.67500	.00005
2.43000	.00020	2.51200	.00013	2.59400	.00008	2.67600	.00005
2.43100	.00020	2.51300	.00013	2.59500	.00008	2.67700	.00005
2.43200	.00020	2.51400	.00012	2.59600	.00008	2.67800	.00005
2.43300	.00020	2.51500	.00012	2.59700	.00008	2.67900	.00005
2.43400	.00020	2.51600	.00012	2.59800	.00008	2.68000	.00005
2.43500	.00020	2.51700	.00012	2.59900	.00008	2.68100	.00005
2.43600	.00019	2.51800	.00012	2.60000	.00008	2.68200	.00005
2.43700	.00019	2.51900	.00012	2.60100	.00008	2.68300	.00005
2.43800	.00019	2.52000	.00012	2.60200	.00007	2.68400	.00005
2.43900	.00019	2.52100	.00012	2.60300	.00007	2.68500	.00005
2.44000	.00019	2.52200	.00012	2.60400	.00007	2.68600	.00005
2.44100	.00019	2.52300	.00012	2.60500	.00007	2.68700	.00004
2.44200	.00019	2.52400	.00012	2.60600	.00007	2.68800	.00004
2.44300	.00019	2.52500	.00012	2.60700	.00007	2.68900	.00004
2.44400	.00019	2.52600	.00012	2.60800	.00007	2.69000	.00004
2.44500	.00018	2.52700	.00012	2.60900	.00007	2.69100	.00004
2.44600	.00018	2.52800	.00011	2.61000	.00007	2.69200	.00004
2.44700	.00018	2.52900	.00011	2.61100	.00007	2.69300	.00004
2.44800	.00018	2.53000	.00011	2.61200	.00007	2.69400	.00004
2.44900	.00018	2.53100	.00011	2.61300	.00007	2.69500	.00004
2.45000	.00018	2.53200	.00011	2.61400	.00007	2.69600	.00004
2.45100	.00018	2.53300	.00011	2.61500	.00007	2.69700	.00004
2.45200	.00018	2.53400	.00011	2.61600	.00007	2.69800	.00004
2.45300	.00018	2.53500	.00011	2.61700	.00007	2.69900	.00004
2.45400	.00018	2.53600	.00011	2.61800	.00007	2.70000	.00004
2.45500	.00017	2.53700	.00011	2.61900	.00007	2.70100	.00004
2.45600	.00017	2.53800	.00011	2.62000	.00007	2.70200	.00004
2.45700	.00017	2.53900	.00011	2.62100	.00007	2.70300	.00004
2.45800	.00017	2.54000	.00011	2.62200	.00007	2.70400	.00004
2.45900	.00017	2.54100	.00011	2.62300	.00007	2.70500	.00004
2.46000	.00017	2.54200	.00011	2.62400	.00007	2.70600	.00004
2.46100	.00017	2.54300	.00011	2.62500	.00007	2.70700	.00004
2.46200	.00017	2.54400	.00010	2.62600	.00006	2.70800	.00004
2.46300	.00017	2.54500	.00010	2.62700	.00006	2.70900	.00004
2.46400	.00017	2.54600	.00010	2.62800	.00006	2.71000	.00004
2.46500	.00016	2.54700	.00010	2.62900	.00006	2.71100	.00004
2.46600	.00016	2.54800	.00010	2.63000	.00006	2.71200	.00004
2.46700	.00016	2.54900	.00010	2.63100	.00006	2.71300	.00004
2.46800	.00016	2.55000	.00010	2.63200	.00006	2.71400	.00004
2.46900	.00016	2.55100	.00010	2.63300	.00006	2.71500	.00004
2.47000	.00016	2.55200	.00010	2.63400	.00006	2.71600	.00004
2.47100	.00016	2.55300	.00010	2.63500	.00006	2.71700	.00004
2.47200	.00016	2.55400	.00010	2.63600	.00006	2.71800	.00004
2.47300	.00016	2.55500	.00010	2.63700	.00006	2.71900	.00004
2.47400	.00016	2.55600	.00010	2.63800	.00006	2.72000	.00004
2.47500	.00016	2.55700	.00010	2.63900	.00006	2.72100	.00004
2.47600	.00015	2.55800	.00010	2.64000	.00006	2.72200	.00004
2.47700	.00015	2.55900	.00010	2.64100	.00006	2.72300	.00004
2.47800	.00015	2.56000	.00010	2.64200	.00006	2.72400	.00004
2.47900	.00015	2.56100	.00009	2.64300	.00006	2.72500	.00004
2.48000	.00015	2.56200	.00009	2.64400	.00006	2.72600	.00004
2.48100	.00015	2.56300	.00009	2.64500	.00006	2.72700	.00004
2.48200	.00015	2.56400	.00009	2.64600	.00006	2.72800	.00003
2.48300	.00015	2.56500	.00009	2.64700	.00006	2.72900	.00003
2.48400	.00015	2.56600	.00009	2.64800	.00006	2.73000	.00003
2.48500	.00015	2.56700	.00009	2.64900	.00006	2.73100	.00003
2.48600	.00015	2.56800	.00009	2.65000	.00006	2.73200	.00003
2.48700	.00015	2.56900	.00009	2.65100	.00006	2.73300	.00003
2.48800	.00014	2.57000	.00009	2.65200	.00006	2.73400	.00003
2.48900	.00014	2.57100	.00009	2.65300	.00006	2.73500	.00003
2.49000	.00014	2.57200	.00009	2.65400	.00005	2.73600	.00003
2.49100	.00014	2.57300	.00009	2.65500	.00005	2.73700	.00003
2.49200	.00014	2.57400	.00009	2.65600	.00005	2.73800	.00003
2.49300	.00014	2.57500	.00009	2.65700	.00005	2.73900	.00003
2.49400	.00014	2.57600	.00009	2.65800	.00005	2.74000	.00003
2.49500	.00014	2.57700	.00009	2.65900	.00005	2.74100	.00003
2.49600	.00014	2.57800	.00009	2.66000	.00005	2.74200	.00003
2.49700	.00014	2.57900	.00009	2.66100	.00005	2.74300	.00003
2.49800	.00014	2.58000	.00008	2.66200	.00005	2.74400	.00003
2.49900	.00014	2.58100	.00008	2.66300	.00005	2.74500	.00003
2.50000	.00014	2.58200	.00008	2.66400	.00005	2.74600	.00003
2.50100	.00013	2.58300	.00008	2.66500	.00005	2.74700	.00003
2.50200	.00013	2.58400	.00008	2.66600	.00005	2.74800	.00003
2.50300	.00013	2.58500	.00008	2.66700	.00005	2.74900	.00003

$\frac{r}{\sqrt{4Tt}}$	$\int_{r/\sqrt{4Tt}}^{\infty}\frac{1}{u}e^{-u^2}du$	$\frac{r}{\sqrt{4Tt}}$	$\int_{r/\sqrt{4Tt}}^{\infty}\frac{1}{u}e^{-u^2}du$	$\frac{r}{\sqrt{4Tt}}$	$\int_{r/\sqrt{4Tt}}^{\infty}\frac{1}{u}e^{-u^2}du$	$\frac{r}{\sqrt{4Tt}}$	$\int_{r/\sqrt{4Tt}}^{\infty}\frac{1}{u}e^{-u^2}du$
2.75000	.00003	2.81300	.00002	2.87600	.00001	2.93900	.00001
2.75100	.00003	2.81400	.00002	2.87700	.00001	2.94000	.00001
2.75200	.00003	2.81500	.00002	2.87800	.00001	2.94100	.00001
2.75300	.00003	2.81600	.00002	2.87900	.00001	2.94200	.00001
2.75400	.00003	2.81700	.00002	2.88000	.00001	2.94300	.00001
2.75500	.00003	2.81800	.00002	2.88100	.00001	2.94400	.00001
2.75600	.00003	2.81900	.00002	2.88200	.00001	2.94500	.00001
2.75700	.00003	2.82000	.00002	2.88300	.00001	2.94600	.00001
2.75800	.00003	2.82100	.00002	2.88400	.00001	2.94700	.00001
2.75900	.00003	2.82200	.00002	2.88500	.00001	2.94800	.00001
2.76000	.00003	2.82300	.00002	2.88600	.00001	2.94900	.00001
2.76100	.00003	2.82400	.00002	2.88700	.00001	2.95000	.00001
2.76200	.00003	2.82500	.00002	2.88800	.00001	2.95100	.00001
2.76300	.00003	2.82600	.00002	2.88900	.00001	2.95200	.00001
2.76400	.00003	2.82700	.00002	2.89000	.00001	2.95300	.00001
2.76500	.00003	2.82800	.00002	2.89100	.00001	2.95400	.00001
2.76600	.00003	2.82900	.00002	2.89200	.00001	2.95500	.00001
2.76700	.00003	2.83000	.00002	2.89300	.00001	2.95600	.00001
2.76800	.00003	2.83100	.00002	2.89400	.00001	2.95700	.00001
2.76900	.00003	2.83200	.00002	2.89500	.00001	2.95800	.00001
2.77000	.00003	2.83300	.00002	2.89600	.00001	2.95900	.00001
2.77100	.00003	2.83400	.00002	2.89700	.00001	2.96000	.00001
2.77200	.00003	2.83500	.00002	2.89800	.00001	2.96100	.00001
2.77300	.00003	2.83600	.00002	2.89900	.00001	2.96200	.00001
2.77400	.00003	2.83700	.00002	2.90000	.00001	2.96300	.00001
2.77500	.00003	2.83800	.00002	2.90100	.00001	2.96400	.00001
2.77600	.00003	2.83900	.00002	2.90200	.00001	2.96500	.00001
2.77700	.00003	2.84000	.00002	2.90300	.00001	2.96600	.00001
2.77800	.00003	2.84100	.00002	2.90400	.00001	2.96700	.00001
2.77900	.00003	2.84200	.00002	2.90500	.00001	2.96800	.00001
2.78000	.00003	2.84300	.00002	2.90600	.00001	2.96900	.00001
2.78100	.00003	2.84400	.00002	2.90700	.00001	2.97000	.00001
2.78200	.00002	2.84500	.00002	2.90800	.00001	2.97100	.00001
2.78300	.00002	2.84600	.00002	2.90900	.00001	2.97200	.00001
2.78400	.00002	2.84700	.00002	2.91000	.00001	2.97300	.00001
2.78500	.00002	2.84800	.00002	2.91100	.00001	2.97400	.00001
2.78600	.00002	2.84900	.00002	2.91200	.00001	2.97500	.00001
2.78700	.00002	2.85000	.00002	2.91300	.00001	2.97600	.00001
2.78800	.00002	2.85100	.00002	2.91400	.00001	2.97700	.00001
2.78900	.00002	2.85200	.00002	2.91500	.00001	2.97800	.00001
2.79000	.00002	2.85300	.00002	2.91600	.00001	2.97900	.00001
2.79100	.00002	2.85400	.00002	2.91700	.00001	2.98000	.00001
2.79200	.00002	2.85500	.00002	2.91800	.00001	2.98100	.00001
2.79300	.00002	2.85600	.00002	2.91900	.00001	2.98200	.00001
2.79400	.00002	2.85700	.00002	2.92000	.00001	2.98300	.00001
2.79500	.00002	2.85800	.00002	2.92100	.00001	2.98400	.00001
2.79600	.00002	2.85900	.00002	2.92200	.00001	2.98500	.00001
2.79700	.00002	2.86000	.00002	2.92300	.00001	2.98600	.00001
2.79800	.00002	2.86100	.00001	2.92400	.00001	2.98700	.00001
2.79900	.00002	2.86200	.00001	2.92500	.00001	2.98800	.00001
2.80000	.00002	2.86300	.00001	2.92600	.00001	2.98900	.00001
2.80100	.00002	2.86400	.00001	2.92700	.00001	2.99000	.00001
2.80200	.00002	2.86500	.00001	2.92800	.00001	2.99100	.00001
2.80300	.00002	2.86600	.00001	2.92900	.00001	2.99200	.00001
2.80400	.00002	2.86700	.00001	2.93000	.00001	2.99300	.00001
2.80500	.00002	2.86800	.00001	2.93100	.00001	2.99400	.00001
2.80600	.00002	2.86900	.00001	2.93200	.00001	2.99500	.00001
2.80700	.00002	2.87000	.00001	2.93300	.00001	2.99600	.00001
2.80800	.00002	2.87100	.00001	2.93400	.00001	2.99700	.00001
2.80900	.00002	2.87200	.00001	2.93500	.00001	2.99800	.00001
2.81000	.00002	2.87300	.00001	2.93600	.00001	2.99900	.00001
2.81100	.00002	2.87400	.00001	2.93700	.00001	3.00000	.00001
2.81200	.00002	2.87500	.00001	2.93800	.00001		

References

ABDUL KHADER, M. H., ELANGO, K., VEERANKUTTY, M. K. and SATYANANDAM, G. Studies on a well penetrating a two aquifer system, *Proc. Int. Symp. on Dev. of Groundwater Resources,* 2, III, 43–52 (1973).

ARNELL, J. C. Permeability studies, surface area measurements using modified Kozeny Equation, *Can. J. Res.,* **A24**, 103–116 (1946).

ARON, C. and SCOTT, V. H. Simplified solution for decreasing flow in wells, *Proc. ASCE HY5,* 1–12 (Sept. 1965).

BACHMAT, Y. On the similitude of dispersion phenomena in homogeneous and isotropic porous medium, *Water Resources Research,* 4, 1079–1083 (1967).

BADON-GHYBEN, W. Nota in verband met de voorgenomen putboring nabij Amsterdam, *Tijdschr. Koninkl. Inst. Ing.,* The Hague (1888).

BAKHMETEFF, B. A. and FEODOROFF, N. V. Flow through granular media, *J. appl. Mech.,* **4A**, 97–104 (1937). (Discussion **5**, 86–90, 1937).

BEAR, J. On the tensor form of dispersion, *J. geophys. Res.,* 66, 1185–1197 (1961 *a*).

BEAR, J. Some experiments on dispersion, *J. geophys. Res.,* 66, 2455–2467 (1961 *b*).

BEAR, J. *Dynamics of fluids in porous media*, American Elsevier, New York (1972).

BEAR, J. and DAGAN, G. The transition zone between fresh and salt waters in a coastal aquifer, Prog. Rep. 1: The steady interface between two immiscible fluids in a two-dimensional field of flow, Hydraulic Lab., Technion, Haifa, IASH (1962).

BEAR, J., ZASLAVSKY, D. and IRMAY, S. *Physical properties of water percolation and seepage,* UNESCO, Paris (1968).

BLACKWELL, R. J., RAYNE, J. R. and TERRY, W. M. Factors influencing the efficiency of miscible displacement, *Trans. AIME* **217**, 1–8 (1959).

BOCKRIS, J. O. M. and REDDY, A. K. N. *Modern electrochemistry* (2 vols), Plenum Press, New York (1970).

BODMAN, G. B. and COLMAN, E. A. Moisture and energy conditions during downward entry of water into soils, *Soil Sci. Soc. Am. Proc.,* **7**, 116–122 (1943).

BOULTON, N. S. The drawdown of the water-table under non-steady conditions near a pumped well in an unconfined formation, *Proc. ICE,* Part 3, 3, 564–579 (1954).

BOULTON, N. S. Analysis of data from non-equilibrium pumping tests allowing for delayed yield from storage, *Proc. ICE,* **26**, 469–482 (1963).

BOULTON. N. S. The discharge to a well in an extensive unconfined aquifer with constant pumping level, *J. Hydrology,* **3**, 124–130 (1965).

BROOKS, R. H. and COREY, A. T. Hydraulic properties of porous media, *Hydrology Papers,* Colorado State University (1964).

BROOKS, R. H. and COREY, A. T. Properties of porous media affecting fluid flow, *Proc, ASCE (J. Irrigation and Drainage Div.),* Vol. 92, No. IR2, 61–88 (1966).

BUCKINGHAM, E. Studies in the movement of soil moisture, *US Dept. Agr. Bur. Soils Bull.,* **38**, 29–61 (1907).

BURDINE, N. T. Relative permeability calculations from pore-size distribution data, *Trans. AIME,* **198**, 71–78 (1953).

CARSLAW, H. S. and JAEGER, J. C. *Conduction of heat in solids* (2nd Ed.), Oxford University Press, London (1959).

CASAGRANDE, A. Seepage through dams, *New England Waterworks Association,* Vol. LI, No. 2, 131–171 (1937).

CEDERGREN, H. R. *Seepage, drainage and flow nets,* Wiley, New York (1967).

CHILDS, E. C. *The physical basis of soil water phenomena,* Wiley-Interscience, London (1969).

CHILDS, E. C. and COLLIS-GEORGE, N. Permeability of porous materials, *Proc. Roy. Soc.,* London, **A201**, 392–405 (1950).

COOPER, H. H. Jr. and JACOB, C. E. A generalized graphical method for evaluating formation constants and summarizing field history, *Trans. Am. Geophys. Union,* **27**, 526–534 (1956).

COOPER, H. H. Jr. and RORABOUGH, M. I. Groundwater movements and bank storage due to flood stages in surface streams, *Geological Survey Water Supply Paper 1536–J,* U.S. Dept. of the Interior Geological Survey, Washington, D.C. (1963).

CRANK, J. *Mathematics of diffusion,* Oxford University Press, London (1956).

DE JONG, J. Longitudinal and transverse diffusion in granular deposits, *Trans. Am. Geophys. Union,* **39**, 67–74 (1958).

DE WIEST, R. J. M. Unsteady flow through an undrained earth dam, *J. Fluid Mech.,* **8**, 1–9 (1960).

DE WIEST, R. J. M. *Geohydrology,* John Wiley, New York (1965).

DUDGEON, C. R., HUYAKORN, P. S. and SWAN, W. H. C. Hydraulics of flow near wells in unconsolidated sediments, Vol. 1, (1972): Theoretical and experimental studies, Vol. 2, (1973), Field Studies, *Water Research Laboratory Report* No. 126, University of New South Wales, Manly Vale, NSW Australia.

EAGLESON, P. S. *Dynamic Hydrology,* McGraw-Hill, New York (1970).

EDLEFSEN, N. E. and ANDERSON A. B. C. Thermodynamics of soil moisture, *Hilgardia,* No. 2, **15**, 31–298 (1943).

ELDER, J. W. Turbulent free convection in a vertical slot, *J. Fluid Mech.,* **23**, 77–98 (1965).

ENGELUND, F. Pumping from leaky artesian aquifers, *Nordic Hydrology,* **1**, 150–157 (1970).

FAIR, G. M. and HATCH, L. P. Fundamental factors governing the streamline flow of water through sand, *J. Am. Wat. Wks. Ass.,* **25**, 1551–1565 (1933).

FORCHHEIMER, P. Wasserbewegung durch Boden, *Z. Ver. Dt. Ing.*, **45** 1782–1788 (1901).

GARDNER, W. R. Some steady state solutions of the unsaturated moisture flow equation with application to evaporation from a water table, *Soil. Sci.*, **85**, 228–232 (1958).

GLOVER, R. E. Ground water movement, *Engineering Monograph No. 31*, (2nd Printing), US Dept. of Interior, Bureau of Reclamation, Denver, Colorado (1966).

HAMMAD, H. Y. Future of groundwater in African Sahara Desert, *Proc. ASCE*, Vol. 95, IR4, 563–580 (Dec. 1969).

HANTUSH, M. S. Analysis of data from pumping tests in leaky aquifers, *Trans. Am. Geophys. Union*, **37**, 702–714 (1956).

HANTUSH, M. S. Analysis of data from pumping wells near a river. *J. geophys. Res.*, **64**, 1921–32 (1959a).

HANTUSH, M. S. Non-steady flow to flowing wells in leaky aquifers. *J. geophys. Res.*, **64**, 1043–1052 (1959b).

HANTUSH, M. S. Modification of theory of leaky aquifers, *J. geophys. Res.*, **65**, 3713–3725 (1960).

HANTUSH, M. S. Drawdown around a partially penetrating well, *Proc. ASCE*, Vol. 86, HY4, 83–98 (July 1961).

HANTUSH, M. S. Flow of groundwater in sands of non-uniform thickness: Part I Flow in a wedge-shaped aquifer, Part II Approximate theory, *J. geophys. Res.*, **67**, 703–720 (1962a).

HANTUSH, M. S. Drainage wells in leaky water-table aquifers, *Proc. ASCE*, Vol. 87, HY2, 123–137 (1962b).

HANTUSH, M. S. Flow of groundwater in sands of non-uniform thickness: Part III – Flow to wells, *J. geophys. Res.*, **67**, 1527–1534 (1962c).

HANTUSH, M. S. Depletion of storage, leakage and river flow by gravity wells in sloping sands, *J. geophys. Res.*, **69**, 2551–2560 (1964).

HANTUSH, M. S. and JACOB, C. E. Plane potential flow of groundwater with linear leakage, *Trans. Am. Geophys. Union*, **35**, 917–936 (1954).

HANTUSH, M. S. and JACOB, C. E. Non-steady radial flow in an infinite leaky aquifer and non-steady Green's Functions for an infinite strip of leaky aquifer, *Trans. Am. Geophys. Union*, **36**, 95–112 (1955).

HERZBERG, A. Die Wasserversorgung einiger Nordseebäder, *J. Gasbeleucht.*, **44**, 815–819, 842–844 (1901).

HOOGHOUDT, S. N. "Bijdragen tot de kennis van eenige natuurkundige grootheden van den ground, 7. Algemeene beschouwing van het problem van de detail ontwatering en de infiltratie door middel van parallel loopende drains, grepples, slooten en kanalen", *Versl. Landbouwk. Ond.*, **46**, 515–707, The Hague (1940).

HUBBERT, M. K. Darcy's Law and the field equations of the flow of underground fluids, *Trans. Am. Inst. Min. Metal. Eng.*, **207**, 222–239 (1956).

HUISMAN, L. *Groundwater recovery*, Macmillan, London (1972).

HUISMAN, L. and KEMPERMAN, J. Bemaling van Spanningsgrondwater, *De Ingenieur*, B. Bouw-en Waterbouwkunde 4, Vol. 62, No. 13, 29–35 (1951).

HUNT, B. W. Exact flow rates from Dupuit's approximation, *Proc. ASCE.*, Vol. 96, No. HY3, 633–642 (1970).

HUYAKORN, P. S. Finite element solution of two-regime flow towards wells, *Water Research Laboratory Report No. 137,* University of New South Wales, Manly Vale, NSW Australia (1973).

HUYAKORN, P. S. and DUDGEON, C. R. Groundwater and well hydraulics: an annotated bibliography, *Water Research Laboratory Report No. 121.,* University of New South Wales, Manly Vale, NSW Australia (1972).

HUYAKORN, P. S. and DUDGEON, C. R. Finite element programs for analysing flow towards wells, *Water Research Laboratory Report No. 135,* University of New South Wales, Manly Vale, NSW Australia (1974).

JACOB, C. E. Radial flow in a leaky artesian aquifer, *Trans. Am. Geophys. Union,* **27**, 198–205 (1946).

JACOB, C. E. Flow of groundwater. In *Engineering Hydraulics* (H. Rouse, Ed.), Wiley, New York (1950).

JACKSON, R. D. Water vapour diffusion in relatively dry soil: I Theoretical considerations and sorption experiments, *Soil Sci. Soc. Amer. Proc.,* **28**, 172–176 (1964). II Desorption experiments, *Soil Sci. Soc. Amer. Proc.,* **28**, 464–466 (1964). III Steady state experiments, *Soil Sci. Soc. Amer. Proc.* **28**, 466–470 (1964).

JACKSON, R. D., REGINATO, R. J. and VAN BAVEL C. H. M. Comparison of measured and hydraulic conductivities of unsaturated soils, *Water Resources Research,* **1**, 375–380 (1965).

JAEGER, C. *Engineering fluid mechanics,* Blackie & Sons, London (1956).

JAHNKE, E. *and* EMDE, F. *Tables of functions,* Dover, New York (1945).

KARPLUS, W. J. *Analog simulation,* McGraw-Hill, New York (1958).

KHOSLA, A. N., BOSE, N. K. and TAYLOR, E. M. *Design of weirs on permeable foundations,* Central Board of Irrigation (India), Publication No. 12, Simla (1936).

KIRKHAM, C. E. *Turbulent flow in porous media,* Dept. Civil Eng., University of Melbourne (1967).

KIRKHAM, D. and POWERS, W. L. *Advanced soil physics,* Wiley-Interscience, New York (1972).

KLINKENBERG, L. J. The permeability of porous media to liquids and gases, *Amer. Petrol. Inst. Drilling Prod. Pract.,* 200–213 (1941).

KLOSE, W. Über die Strömung verdünnter Gase durch Kapillaren. *Annln Phys.,* **11**, 73–93 (1931).

KNUDSEN, M. *The kinetic theory of gases* (3rd Ed.), Methuen, London (1950).

KOZENY, J. Grundwasserbewegung bei freiem Spiegel, Fluss- und Kanalversicherung, *Wasserkraft und Wasserwirtschaft No. 3* (1931).

KRAIJENHOFF VAN DE LEUR, D. A. A study of non-steady groundwater flow with special reference to a reservoir coefficient, *de Ingenieur,* Bouw-en Waterbouwkunde, **9**, 87–94, (in English) (1958).

KRUMBEIN, W. C. and MONK, G. D. Permeability as a function of the size parameters of unconsolidated sand, *Am. Inst. Min. & Met. Eng. Tech. Pub.,* 1492, 11 (1942).
Also *Trans. AIME* **151**, 153- (1942).

KUTILEK, M. Non-Darcian flow of water in soils (laminar region), *1st IAHR Symp:* Fundamentals of Transport Phenomena in Porous Media, Haifa, Israel (1969).

LELIAVSKY, S. *Irrigation and hydraulic design,* Vol. 1, Chapman and Hall, London (1955).

LEVERETT, M. C. Capillary behaviour in porous media, *Trans. AIME,* **142,** 341–358 (1941).

MAHMOND, A. A–Z. and SCOTT, V. H. Non-steady flow for wells with decreasing discharge, *Proc. ASCE,* Vol. 88, HY3, 119–132 (May 1963).

MAHMOND, A. A–Z., SCOTT, V. H. and ARON, C. Modified solutions for decreasing discharge wells., *Proc. ASCE,* Vol. 89, HY6, 145–160 (Nov. 1964).

MEINZER, O. E. The history and development of groundwater hydrology, *J. Wash. Acad. Sci.,* **24,** 6–32 (1934).

MEINZER, O. E. *Hydrology,* McGraw-Hill, New York (1942).

MOORE, R. E. Water conduction from shallow water tables, *Hilgardia,* **12,** 383–426 (1939).

MUSKAT, M. *The flow of homogeneous fluids through porous media,* McGraw-Hill, New York (1937).

Netherlands Hydrological Colloquium: Steady flow of groundwater towards wells, *Commit. for Hydrol. Res. TNO, Proc. and Inform.* No. 10, 179, The Hague (1964).

NIKOLAEVSKII, V. N. Convective diffusion in porous media, *J. Appl. Math. Mech. (P.M.M.),* **23,** 1042–1050) (1959).

NGUYEN, VAN U'U *Moisture movement in road basecourses due to temperature gradients,* Ph. D. Thesis, University of Auckland, New Zealand (1974).

ORCHISTON, H. D. Adsorption of water vapour, I. Soils at 25C, *Soil Sci.,* **76,** 453–465 (1953).

PAPADOPULOS, I. S. Non-steady flow to multiaquifer wells, *J. geophys. Res.,* **71,** 4791–4797 (1966).

PFANNKUCH, H. O. Contribution à l'étude des déplacements de fluides miscibles dans un milieu poreux, *Revue de l'Institute Français du Pétrole,* **18,** 215–270 (1963).

PHILIP, J. R. The theory of infiltration, I The infiltration equation and its solution, *Soil Sci.,* **83,** 345–357 (1956).

PHILIP, J. R. Evaporation and moisture and heat fields in the soil, *J. Met.,* **14,** 354–366 (1957).

PHILIP, J. R. Theory of infiltration. In *Advances in Hydroscience* (V.T. Chow Ed.), 215–296, Academic Press, New York (1969).

POLUBARINOVA-KOCHINA, P. YA. *Theory of underground water movement,* Princeton University Press, Princeton (1962).

RAUDKIVI. A. J. Roxburgh Power Project: electric analogy model tests, *Ministry of Works Hydraulics Laboratory Report* No. 24 (1954).

RAUDKIVI. A. J. *Loose Boundary Hydraulics* (2nd Ed.), Pergamon Press, Oxford (1976).

RAUDKIVI, A. J. and CALLANDER, R. A. *Advanced fluid mechanics,* Edward Arnold, London (1975).

RAUDKIVI, A. J. and NGUYEN, VAN U'U Moisture movement in road base-course due to temperature gradients, *N.Z. Roading Symp* (1975).

RICHARDS, L. A. Capillary conduction of liquids through porous medium, *Physics,* **1,** 318–333 (1931).

ROGERS, J. S. and KLUTE, A. The hydraulic conductivity-water content relationship during non-steady flow through a sand column, *Soi. Sci. Am. Soc. Proc.,* 35, 695–700 (1971).

ROSE, D. A. Determination of capillary conductivity and diffusivity, *Proc. Wageningen Symp.*: "Water in the Unsaturated Zone", UNESCO/IASH, 171–181 (1969).

ROSE, H. E. (1) An investigation into the laws of flow of fluids through beds of granular material. (2) The isothermal flow of gases through beds of granular materials. (3) On the resistance coefficient – Reynolds number relationship for fluid flow through a bed of granular material. *Proc. Inst. Mech. Eng.,* **153**, 141–161 (1945).

ROWE, P. W. Anchored sheet-pile walls, *Proc. ICE,* 1(1), 27 (1952).

ROWE, P. W. A stress-strain theory for cohesionless soil with application to earth pressures at rest and moving walls, *Geotechnique,* Vol. IV, No. 2, 70 (1954).

ROWE, P. W. A theoretical and experimental analysis of sheet-pile walls, *Proc. ICE,* Vol. 4, No. 1, 32 (1955).

ROWE, P. W. Sheet-pile walls in clay, *Proc. ICE,* Vol. 7 (July), 629 (1957).

ROWE, P. W. Stress dilatancy, earth pressures and slopes, *Proc. ASCE, J. Soil Mechanics and Foundations Div.,* 37–61 (May 1963).

SAFFMAN, P. G. A theory of dispersion in a porous medium, *J. Fluid Mech.,* 6, 321–349 (1959).

SAFFMAN, P. G. Dispersion due to molecular diffusion and macroscopic mixing in flow through a network of capillares, *J. Fluid Mech.,* 7, 194–208 (1960).

SCHAFFERNAK, F. Über die Stausicherheit durchlässiger geschütteter Dämme, *Allg. Bauztg* (1917).

SCHEIDEGGER, A. E. *The physics of flow through porous media,* (2nd Ed.), University of Toronto Press, Toronto (1960).

SHERARD, J. L., WOODWARD, R. J., GIZIENSKI, S. F. and CLEVENGER, W. A. *Earth and rock fill dams: engineering problems of design and construction,* Wiley, New York (1963).

SKAGGS, R. W., MONKE, E. J. and HUGGINS, L. F. *An approximate method for determining the hydraulic conductivity function of unsaturated soil,* Purdue University Water Resources Research Center, Lafayette, Indiana (1970).

SOKOL, D. Position and fluctuations of water level in wells perforated in more than one aquifer, *J. geophys. Res.,* **68**, 1079–1080 (1963).

STERNBERG, Y. M. Some approximate solutions of radial flow problems, *J. Hydrology,* 7, 158–166 (1969).

STERNBER, Y. M. Theory and application of the skin effect concept to groundwater wells, *Proc. Int. Symp. on Dev. of Groundwater Resources,* 2, III 23–32, Madras, India (1973).

SUNADA, D. *Turbulent flow through porous media,* Water Resources Center, Contribution No. 103, University of California, Berkeley (1965).

SWARTZENDRUBER D. Non-Darcy flow behaviour in liquid-saturated porous media, *J. geophys. Res.* No. 13, **67**, 5205–5213 (1962).

SWARTZENDRUBER, D. Comment on the paper: Determination of the hydraulic conductivity of unsaturated soils from an analysis of transient flow data, by G. Vachaud (1967), *Water Resources Res.* No. 3, **4**, 659–660 (1968).

SWARTZENDRUBER, D. Chapter 6 of de Weist, R.J.M.(Ed.) *Flow through porous media,* Academic Press, New York (1969).

THEISS, C. V. The relation between the lowering of the piezometric surface and the rate and duration of discharge of a well using groundwater storage, *Trans. Am. Geophys. Union,* **16**, 519–524 (1935).

TILLER, F. M. Role of porosity in filtration, *Chem. Engng. Prog.,* **49**, 467–479 (1953).

TILLER, F. M. Role of porosity in filtration, *Chem. Engng. Prog.,* **51**, 282–290 (1955).

TISON, G. Jr. Fluctuations of groundwater levels. In *Advances in Geophysics,* **11**, 303–326, Edited by H. E. Landsberg and J. van Mieghem, Academic Press (1965).

VAN DER LEEDEN, F. (compiler). *Ground water; A Selected Bibliography* (2nd Ed). Water Information Center Inc., Port Washington, N.Y. (1974).

VAN SCHILFGAARDE, J., KIRKHAM, D. and FREVERT, R. K. *Physical and mathematical theories of tile and ditch drainage and their usefulness in design,* Agricultural Experiment Station, Iowa State College, Research Bulletin, 436 (1956).

VERRUIJT, A. *Theory of groundwater flow,* Macmillan, London (1970).

WALSHAW, A. C. *Engineering units and worked examples for students of the mechanical sciences* (2nd Ed.), Blackie, London (1964).

WALTON, W. C. *Groundwater resource evaluation,* McGraw-Hill, New York (1970).

WENZEL, L. K. and FISHEL, V. C. Methods for determining permeability of water-bearing materials with special reference to discharging well methods, *U.S. Geological Survey Water-Supply Paper,* 887, Washington, D.C. (1942).

WOODING, R. A. Convection in a saturated porous medium at large Rayleigh number or Peclet number, *J. Fluid Mech.,* **15**, 527-544 (1963).

WOODING, R. A. Mixing-layer flows in a saturated porous medium, *J. Fluid Mech.,* **19**, 103–112 (1964).

WOODING, R. A. and CHAPMAN, T. G. Groundwater flow over a sloping impermeable layer, 1 Application of the Dupuit-Forchheimer assumption, *J. geophys. Res.,* **71**, 2895–2902 (1966).

Index